ALLIED MATHEMATICS I & II
QUESTION BANK

MJP
PUBLISHERS

ALLIED MATHEMATICS I & II
QUESTION BANK

K. MAHABOOB HASSAIN SHERIEFF M.Sc., M.Phil., Ph.D.,

Assistant Professor of Mathematics

C. Abdul Hakeem College (Autonomous),

Melvisharam – 632509.

Ranipet District.

ISBN 9789355287090   MJP Publishers

All rights reserved   No. 44, Nallathambi Street,
Printed and bound in India   Triplicane, Chennai 600 005

MJP 1659   Publishers, 2024

Publisher :   C. Janarthanan

Project Editor :   C. Ambica

PREFACE

This book has been written in accordance with the latest syllabus of Thiruvalluvar University, Tamilnadu, for the first and second year B.Sc., (Physics/chemistry) students.

This book contains 2marks, 5 marks and 10 marks problems with solutions for each units in Allied Mathematics I and Allied Mathematics II.

A large number of exercise problems and solutions have been added in each chapter to enable the students to learn at their own pace.

This book is aimed to help first year and second year allied mathematics students to secure very high marks in mathematics.

I have tried my best to make this book free from errors, omissions and discrepancies.

Suggestions for the improvement of this book are most welcome.

I am really thankful to my publisher for showing personal interest and bringing out this book in a short period.

Dr. K. Mahaboob Hassain Sherieff (AUTHOR)

Assistant Professor in Mathematics,

C. Abdul Hakeem College (Autonomous),
Melvisharam- 632509.
Ranipet District.

CONTENTS

ALLIED MATHEMATICS – I

ALGEBRA

2 MARKS

1) Show that $\dfrac{\dfrac{1}{2!}+\dfrac{1}{4!}+\dfrac{1}{6!}+...}{\dfrac{1}{1!}+\dfrac{1}{3!}+\dfrac{1}{5!}+...}=\dfrac{e-1}{e+1}$

Sol: LHS $=\dfrac{\dfrac{1}{2!}+\dfrac{1}{4!}+\dfrac{1}{6!}+...}{\dfrac{1}{1!}+\dfrac{1}{3!}+\dfrac{1}{5!}+...}=\dfrac{\dfrac{e+e^{-1}}{2}-1}{\dfrac{e-e^{-1}}{2}}=\dfrac{e-e^{-1}-2}{e-e^{-1}}=\dfrac{e^2+1-2e}{e^2-1}$

$\qquad\qquad =\dfrac{\left(e-1\right)^2}{\left(e+1\right)\left(e-1\right)}=\dfrac{e-1}{e+1}=\text{RHS}$

2) Prove that $\dfrac{a-x}{a}+\dfrac{1}{2}(\dfrac{a-x}{a})^2+\dfrac{1}{3}(\dfrac{a-x}{a})^2+...=\mathbf{\log a-\log_e X}.$

Sol: LHS $=\dfrac{a-x}{a}+\dfrac{1}{2}(\dfrac{a-x}{a})^2+\dfrac{1}{3}(\dfrac{a-x}{a})^3+...$

$\qquad\qquad =y+\dfrac{y^2}{2}+\dfrac{y^3}{2}+...,\quad \text{where } y=\dfrac{a-x}{a}$

$\qquad\qquad =-\log(1-y)$

$\qquad\qquad =-\log(1-\dfrac{a-x}{a})$

$$= -\log\left(\frac{x}{a}\right)$$

$$= -(\log x - \log a)$$

$$= \log a - \log x \quad = \text{RHS}$$

3) Write the expansion of $(1 + 4x)^{-1/2}$

Sol: We know that

$$(1+x)^{\frac{p}{q}} = 1 - \frac{p}{1!}\left[\frac{x}{q}\right] + \frac{p(p+q)}{2!}\left[\frac{x}{q}\right]^2 - \frac{p(p+q)(p+2q)}{3!}\left[\frac{x}{q}\right]^3 + \ldots$$

Here x = 4x, p = 1, q = 2.

Therefore,

$$(1+4x)^{-1/2} = 1 - \frac{1}{1!}\left[\frac{4x}{2}\right] + \frac{1(1+2)}{2!}\left[\frac{4x}{2}\right]^2 - \frac{1(1+2)(1+2(2))}{3!}\left[\frac{4x}{2}\right]^3 + \ldots$$

$$= 1 - \frac{1}{1!}[2x] + \frac{1.3}{2!}[2x]^2 - \frac{1.3.5}{3!}[2x]^3 + \ldots$$

4) Find the coefficient of x^n in the expansion of $e^{(a+bx)}$

Sol: Given, $e^{(a+bx)} = e^a \cdot e^{bx}$

$$= e^a\left[1 + \frac{bx}{1!} + \frac{(bx)^2}{1!} + \ldots + \frac{(bx)^n}{n!} + \ldots \infty\right]$$

Therefore, the coefficient of xn in this expansion $= \dfrac{e^a b^n}{n!}$

5) Write the expansion of $(1-x)^{-3}$.

Sol: $(1-x)^{-3}\dfrac{1}{2}\left[1.2 + 2.3x + 3.4x^2 + \ldots + (n+1)(n+2)x^n + \ldots \infty\right]$

6) Expand ax in ascending powers of x, a being positive.

Sol: Given, $a^x = e^{x \log a}$

$$= 1 + \frac{x \log a}{1!} + \frac{x^2 (\log a)^2}{2!} + \dots + \frac{x^n (\log a)^n}{n!} + \dots \infty$$

7) Find the expansion of $(1-x)^{-2}$.

Sol: $(1-x)^{-2} = 1 + 2x + 3x^2 + 4x^3 + \dots + (n + 1)x^n + \dots \infty$

8) Resolve into partial fractions $\dfrac{1}{x^2 + 3x + 2}$.

Sol: $= \dfrac{1}{x^2 + 3x + 2} = \dfrac{A}{(x+1)} + \dfrac{B}{(x+2)}$

$$1 = A(x+2) + B(x+2)$$

Put x = -1, we get A = 1

Put x = -2, we get B = -1

$$\frac{1}{x^2 + 3x + 2} = \frac{1}{(x+1)} + \frac{1}{(x+2)}.$$

9) Sum to infinity the series $1 + \dfrac{1+2}{2!} + \dfrac{1+2+2^2}{3!} + \ddot{u} \; \infty$

Sol: The nth term of the series is $T_n = \dfrac{1 + 2 + 2^2 + \dots + 2^{n-1}}{n!}$

$$= \frac{2^{n-1}}{n!}$$

$$= \frac{2^n}{n!} - \frac{1}{n!}$$

Putting n = 1, 2, 3, … we get $T_1 = \dfrac{2}{1!} - \dfrac{1}{1!}$

$$T_2 = \frac{2^2}{2!} - \frac{1}{2!}$$

$$T_3 = \frac{2^3}{3!} - \frac{1}{3!}$$

Adding all the terms, we get

$$S_\infty = \left[\frac{2}{1!} + \frac{2^2}{2!} + \frac{2^3}{3!} + ... + \infty\right] - \left[\frac{1}{1!} + \frac{1}{2!} + \frac{1}{3!} + ... + \infty\right]$$

10) Resolve into partial fractions $\dfrac{x}{(x-1)(x-2)}$

Sol: $\dfrac{x}{(x-1)(x-2)} = \dfrac{A}{(x-1)} + \dfrac{B}{(x-2)}$

$$x = A(x-2) + B(x-1)$$

Put x=1, we get A = -1

Put x=2, we get B = 2

$$\frac{x}{(x-1)(x-2)} = \frac{-1}{(x-1)} + \frac{2}{(x-2)} = \frac{2}{(x-2)} - \frac{-1}{(x-1)}$$

11) Write the expansion of log(1+x) and log(1-x).

Sol: $\log(1+x) = x - \dfrac{x^2}{2} + \dfrac{x^3}{3} + \dfrac{x^4}{4} + ... + (-1)^{n-1}(\dfrac{x^n}{n}) + ..$

$$\log(1+x) = -[x + \frac{x^2}{2} + \frac{x^3}{3} + \frac{x^4}{4} + ...$$

12) show that $\log 10 = 3\log 2 + \dfrac{1}{4} - \dfrac{1}{2}\cdot\dfrac{1}{4^2} + \dfrac{1}{3}\cdot\dfrac{1}{4^3} \ldots \infty$

Sol: $3\log 2 + \dfrac{1}{4} - \dfrac{1}{2}\cdot\dfrac{1}{4^2} + \dfrac{1}{3}\cdot\dfrac{1}{4^3} - \ldots \infty = 3\log 2 + x - \dfrac{x^2}{2} + \dfrac{x^4}{4} - \ldots,$

$$= 3\log 2 + \log(1 + x)$$

$$= 3\log 2 + \log\left(1 + \dfrac{1}{4}\right)$$

$$= 3\log 2 + \log\left(\dfrac{5}{4}\right)$$

$$= \log 8 + \log\left(\dfrac{5}{4}\right)$$

$$= \log\left(8 \times \dfrac{5}{4}\right)$$

$$= \log_e 10.$$

5 MARKS

1) Resolve $\dfrac{7x-1}{1-5x-6x^2}$ into partial fractions.

Sol: Let $\dfrac{7x-1}{1-5x-6x^2} = \dfrac{1-7x}{6x^2+5x-1} = \dfrac{A}{x+1} + \dfrac{B}{6x-1)}$

$$1 - 7x = A(6x - 1) + B(x + 1)$$

Put $x = 1$, *we get* $A = \dfrac{-8}{7}$

Put $x = \dfrac{1}{6}$ *we get* $B = \dfrac{-1}{7}$

Therefore, $\dfrac{7x-1}{1-5x-6x^2} = \dfrac{-8}{7(x+1)} - \dfrac{1}{7(6x-1)}$

2) Show that $\dfrac{1}{1.2} - \dfrac{1}{2.3} + \dfrac{1}{3.4} - \dfrac{1}{4.5} + \ldots = 2\log a\, 2 - 1$

Sol: General Term, $t_n = (-1)^{n+1}\left(\dfrac{1}{n(n+1)}\right), n = 1, 2, 3, \ldots$

Now $\dfrac{1}{n(n+1)} = \dfrac{A}{n} + \dfrac{B}{n+1}$

$$1 = A(n+1) + Bn$$

$$n = 0 \;\Rightarrow\; A = 1$$

$$n = -1 \;\Rightarrow\; B = -1$$

$$t_n = (-1)^{n+1}\left(\dfrac{1}{n} - \dfrac{1}{n+1}\right), n = 1, 2, 3, \ldots$$

$$t_n = \dfrac{(-1)^{n+1}}{n} + \dfrac{(-1)^n}{n+1}$$

$$t_1 = \dfrac{1}{1} - \dfrac{1}{2}$$

$$t_2 = \dfrac{1}{2} + \dfrac{1}{3}$$

$$t_3 = \dfrac{1}{3} + \dfrac{1}{4}$$

$$t_4 = \dfrac{1}{4} + \dfrac{1}{5}$$

$$\ldots \quad \ldots \quad \ldots \quad \ldots \quad \ldots \quad \ldots$$

Therefore, $S = 1 - 2.\dfrac{1}{2} + 2.\dfrac{1}{3} - 2.\dfrac{1}{4} + \ldots$

$$= 1 + 2\left(-\frac{1}{2} + \frac{1}{3} - \frac{1}{4} + \ldots\right)$$

$$= 1 + 2(\log 2 - 1)$$

$$= 1 + 2\log 2 - 2$$

$$= 2\log 2 - 1.$$

3) Resolve into partial fractions $\dfrac{x^3}{(x+a)(x+b)(x+c)}$ Hence deduce that $\displaystyle\sum \dfrac{a^3}{(a-b)(a-c)(a-d)} = 1$.

Sol: Let $\dfrac{x^3}{(x-a)(x-b)(x-c)} = 1 + \dfrac{R}{(x-a)(x-b)(x-c)}$

Where R is the remainder and is of degree less than 3.

Therefore, $\dfrac{x^3}{(x-a)(x-b)(x-c)} = 1 + \dfrac{A}{(x-a)} + \dfrac{B}{(x-b)} + \dfrac{C}{(x-c)}$

$X^3 = (x–a)(x–b)(x–c) + A(x–b)(x–c) + B((x–a)(x–c) + C(x–a)(x–b)$

Put $x = a : a^3 = A(a-b)(a-c) \qquad A = \dfrac{a^3}{(a-b)(a-c)}$

Similarly $B = \dfrac{b^3}{(b-a)(b-c)}$ and $C = \dfrac{c^3}{(c-a)(c-b)}$

Therefore,

$$\frac{\ddot{u}^3}{(x-a)(x-b)(x-c)} = 1 + \frac{a^3}{(a-b)(a-c)(x-a)} + \frac{b^3}{(b-a)(b-c)(x-b)}$$

$$+ \frac{c^3}{(c-a)(c-b)(x-c)}$$

Put

$$x = d : \frac{\ddot{u}^{\ddot{u}}}{(d-a)(d-b)(d-c)} = 1 + \frac{}{(a-b)(a-c)(d-a)} + \frac{}{(b-a)(b-c)(d-b)}$$

$$+ \frac{c^3}{(c-a)(c-b)(d-c)}$$

Therefore, $\Sigma \dfrac{a^3}{(a-b)(a-c)(a-d)}$

4) Prove that $\log \dfrac{n+1}{n-1} = \dfrac{2n}{n^2+1} + \dfrac{1}{3}\left(\dfrac{2n}{n^2+1}\right)^3 + \dfrac{1}{5}\left(\dfrac{2n}{n^2+1}\right)^5 + \cdots$

Sol: $\text{RHS} = \dfrac{2n}{n^2+1} + \dfrac{1}{3}\left(\dfrac{2n}{n^2+1}\right)^3 + \dfrac{1}{5}\left(\dfrac{2n}{n^2+1}\right)^5 + \cdots$

$$= y + \frac{y^3}{3} + \frac{y^5}{5} + \cdots \text{ where } y = \frac{2n}{n^2+1}$$

$$= \frac{1}{2} \log \left(\frac{y}{y}\right)$$

$$= \frac{1}{2} \log \left(\dfrac{1 + \dfrac{2n}{n^2+1}}{1 - \dfrac{2n}{n^2+1}}\right)$$

$$= \frac{1}{2} \log \left(\dfrac{n+1}{n-1}\right)^2$$

$$= \log \left(\dfrac{n+1}{n-1}\right) = \text{LHS}$$

5) Find the sum to infinity to the series $1 + \dfrac{2}{6} + \dfrac{2.5}{6.12} + \dfrac{2.5.8}{6.12.18} + \cdots$

Sol: Let

$$S = 1 + \frac{2}{6} + \frac{2.5}{6.12} + \frac{2.5.8}{6.12.18} + \cdots = 1 + \frac{2}{1!}\left(\frac{1}{6}\right) + \frac{2.5}{2!}(\frac{1}{6})^2 + \frac{2.5.8}{3!}(\frac{1}{6})^2 + \cdots$$

We Know That

$$(1-x)^{\frac{-p}{q}} = 1 + \frac{p}{1!}\left(\frac{x}{q}\right) + \frac{p(p+q)}{2!}(\frac{x}{q})^2 + \frac{p(p+q)(p+2q)}{3!}(\frac{x}{q})^3 + \cdots \;\blacktriangleright\; 1$$

Comparing 1 with the above equation, we get

$$p = 2,\, p+q = 5,\, \frac{x}{q} = \frac{1}{6} \;\blacktriangleright\; p = 2,\, q = 3,\, x = \frac{1}{2}$$

Therefore, $S = (1 - \dfrac{1}{2})^{-2/3} = (\dfrac{1}{2})^{-2/3} = (2)^{2/3}$.

6) Resolve into partial fractions $\dfrac{2x+3}{(x^2+1)(x+4)}$.

Sol: Let $\dfrac{2x+3}{(x^2+1)(x+4)} = \dfrac{A}{x+4} + \dfrac{Bx+c}{x^2+1}$

$$2x + 3 = A(x^2+1) + (Bx+C)(x+4)$$

Put $X = -4 : A = \dfrac{-5}{17}$

Equating the coefficient of x^2, $0 = A+B$ $\blacktriangleright$ $B = \dfrac{5}{17}$

Equating the constant terms, $3 = A+4C$ $\blacktriangleright$ $C = \dfrac{14}{17}$

Therefore, $\dfrac{2x+3}{(x2+1)(x+4)} = \dfrac{-5}{17(x+4)} + \dfrac{5x+14}{17(x2+1)}$.

7) Resolve into partial fractions $\dfrac{x^2}{(x^2+1)(x^2+2)(x^2+3)}$.

Sol: Let $x^2 = y$

Let $\dfrac{y}{(y+1)(y+2)(y+3)} = \dfrac{A}{y+1} + \dfrac{B}{y+2} + \dfrac{C}{y+3}$

$y = A(y + 2)(y + 3) + B(y + 1)(y + 3) + C(y + 1)(y + 2)$

Put $y = -1 : A = \dfrac{-1}{2}$

Put $y = -2 : B = 2$

Put $y = -3 : C = \dfrac{3}{2}$

Therefore, $\dfrac{y}{(y+1)(y+2)(y+3)} = \dfrac{-1}{2(y+1)} + \dfrac{2}{y+2} + \dfrac{3}{2(y+3)}$

i.e., $\dfrac{x^2}{(x^2+1)(x^2+2)(x^2+3)} = \dfrac{-1}{2(x^2+1)} + \dfrac{2}{x^2+2} + \dfrac{3}{2(x^2+3)}$.

8) Show that $\log\left[\dfrac{1+2e^x}{3}\right] = \dfrac{2x}{3} + \dfrac{x^2}{9}$ approximately.

Sol: $\log\left[\dfrac{1+2e^x}{3}\right] = \log\left[\dfrac{1+2\left[1+\dfrac{x}{1!}+\dfrac{x2}{2!}+\cdots\right]}{3}\right]$

$= \log\left[1+\dfrac{2x}{3}+\dfrac{x^2}{3}+\cdots\right]$

$= \left[\dfrac{2x}{3}+\dfrac{x^2}{3}\right] - \dfrac{\left[\dfrac{2x}{3}+\dfrac{x^2}{3}\right]^2}{2}$

$= \dfrac{2x}{3}+\dfrac{x^2}{3}-\dfrac{2x^2}{9}$

$= \dfrac{2x}{3}+\dfrac{x^2}{9}$ approximately.

9) If x is large, then show that $\sqrt{x^2+4}-\sqrt{x^2+1}=\dfrac{3}{2x}+\dfrac{15}{x^3}+\dfrac{63}{16x^5}$

Sol: $\sqrt{x^2+4}-\sqrt{x^2+1}=x\sqrt{1+\dfrac{4}{x^2}}-x\sqrt{1+\dfrac{1}{x^2}}$

$$=x\left[1+\dfrac{4}{x^2}\right]^{\frac{1}{2}}-x\left[1+\dfrac{1}{x^2}\right]^{\frac{1}{2}}$$

$$x\left[1+\dfrac{4}{x^2}\right]^{\frac{1}{2}}=x\left[1+\dfrac{1}{2}\cdot\dfrac{4}{x^2}+\dfrac{\dfrac{1}{2}\left[\dfrac{-1}{2}\right]}{2}\cdot\dfrac{16}{x^4}+\dfrac{\dfrac{1}{2}\left[\dfrac{-1}{2}\right]\left[\dfrac{-3}{2}\right]}{3!}\cdot\dfrac{64}{x^6}+\cdots\right]$$

$$=x+\dfrac{2}{x}-\dfrac{2}{x^3}+\dfrac{4}{x^5}\;\rightarrow\qquad(1)$$

$$x\left[1+\dfrac{1}{x2}\right]\dfrac{1}{2}=x\left[1+\dfrac{1}{2x^2}+\dfrac{\dfrac{1}{2}\left[\dfrac{-1}{2}\right]}{2!}\cdot\dfrac{1}{x4}+\dfrac{\dfrac{1}{2}\left[\dfrac{-1}{2}\right]\left[\dfrac{-3}{2}\right]}{3!}\cdot\dfrac{1}{x6}+\cdots\right]$$

$$=x+\dfrac{1}{2x}-\dfrac{1}{8x^3}+\dfrac{1}{16x^6}\;\rightarrow\qquad(1)$$

Subtracting (2) from (1), $x\left[1+\dfrac{4}{x^2}\right]^{\frac{1}{2}}-x\left[1+\dfrac{1}{x^2}\right]^{\frac{1}{2}}=\dfrac{3}{2x}-\dfrac{15}{8x^3}+\dfrac{63}{16x^5}$.

10) Resolve into partial fractions $\dfrac{2x+3}{(x-1)(x-2)(x-3)}$..

Sol: Let $\dfrac{2x+3}{(x-1)(x-2)(x-3)}=\dfrac{A}{x-1}+\dfrac{B}{x-2}+\dfrac{C}{x-2}$

$$2x + 3 = A(x-2)(x-3) + B(x-1)(x-3) + C(x-1)(x-2)$$

Put $x=1:A=\dfrac{5}{2}$

Put x = 2 : B = − 2

Put $x = 3 : C = \dfrac{9}{2}$

Therefore, $\dfrac{2x+3}{(x-1)(x-2)(x-3)} = \dfrac{5}{2(x-1)} - \dfrac{2}{x-2} + \dfrac{9}{2(x-3)}.$

11) Sum the series $\dfrac{2^2}{1!} + \dfrac{3^2}{2!} + \dfrac{4^2}{3!} + \dfrac{5^2}{4!} + \cdots$

Sol: Let $S = \dfrac{2^2}{1!} + \dfrac{3^2}{2!} + \dfrac{4^2}{3!} + \dfrac{5^2}{4!} + \cdots$

General Term $t_n = \dfrac{(n+1)^2}{n!}, n = 1, 2, 3, \cdots$

Therefore, $S = t_1 + t_2 + t_3 + \ldots$

Let $(n + 1)^2 = A + Bn + Cn(n-1)$

$n = 0 \Rightarrow A = 1$

$n = 1 \Rightarrow A + B = 4 \Rightarrow B = 3$

Equating the coefficient of n^2, we get $C = 1$

Therefore, $(n + 1) = 1 + 3n + n(n - 1)$

$$t_n = \dfrac{1 + 3n + n(n-1)}{n!}$$

$$= \dfrac{1}{n!} + 3\dfrac{n}{n!} + \dfrac{n(n-1)}{n!} = \dfrac{1}{n!} + 3\dfrac{1}{(n-1)!} + \dfrac{1}{(n-2)!}$$

$$t_1 = \dfrac{1}{1!} + 3$$

$$t_2 = \dfrac{1}{2!} + 3\left(\dfrac{1}{1!}\right) + 1$$

$$t_3 = \frac{1}{3!} + 3\left(\frac{1}{2!}\right) + \frac{1}{1!}$$

$$t_4 = \frac{1}{4!} + 3\left(\frac{1}{3!}\right) + \frac{1}{2!}$$

$$\cdots \cdots \cdots \cdots \cdots \cdots \cdots \cdots \cdots \cdots$$

Therefore, $S = \left[\frac{1}{1!} + \frac{1}{2!} + \frac{1}{3!} + \frac{1}{4!} + \cdots\right] + 3\left[1 + \frac{1}{1!} + \frac{1}{2!} + \frac{1}{3!} + \cdots\right] + \left[1 + \frac{1}{1!} + \frac{1}{2!} + \cdots\right]$

$= e - 1 + 3e + e \qquad\qquad = 5e - 1.$

10 MARKS

---

1) Sum to infinity the series $\Sigma \dfrac{n^2}{(n+1)(n+2)} x^n$.

Sol: Let $\dfrac{n^2}{(n+1)(n+2)} = 1 + \dfrac{A}{n+1} + \dfrac{B}{n+2}$

$n^2 = (n+1)(n+2) + A(n+2) + B(n+1)$

Put $n = -1 : A = 1$

Put $n = -2 : B = -4$

$$T_n = \left[1 + \frac{1}{n+1} - \frac{4}{n+2}\right] x^n$$

$$= x^n + \frac{x^n}{n+1} - \frac{x^n}{n+2} = x^n + \frac{1}{x}\frac{x^{n+1}}{n+1} - \frac{4}{x^2}\frac{x^{n+2}}{n+2}$$

$$S = \sum T_n$$

$$= \sum_{n=1}^{\infty} x^n + \frac{1}{x}\sum_{n=1}^{\infty} \frac{x^{n+1}}{n+1} - \frac{4}{x^2}\sum_{n=1}^{\infty} \frac{x^{n+2}}{n+2}$$

$$= (x + x^2 + x^3 + \cdots \infty) + \frac{1}{x}\left[\frac{x^2}{2} + \frac{x^3}{3} + \cdots \infty\right] - \frac{4}{x^2}\left[\frac{x^3}{3} + \frac{x^4}{4} + \frac{x^5}{5} + \cdots \infty\right]$$

$$= \frac{x}{1-x} + \frac{1}{x}[-\log(1-x)-x] - \frac{4}{x^2}\left[-\log(1-x)-x-\frac{x^2}{2}\right]$$

$$= \left[\frac{4}{x^2} - \frac{1}{x}\right]\log(1-x) + \frac{x}{1-x} + \frac{4}{x} + 1.$$

2) Sum to infinity the series $1 + \frac{2^4}{2!} + \frac{3^4}{3!} + \frac{4^4}{4!} + \cdots$

Sol: The nth of the series is $T_n = \frac{n4}{n!} = \frac{n3}{(n-1)!}$

Let $n^3 = A + B(n-1) + C(n-1)(n-2) + D(n-1)(n-2)(n-3)$

Put n = 1, A = 1;

Put n = 2, A + B = 8 ➜ B = 7;

Put n = 3, A + 2B + 2C = 27 ➜ C = 6;

Equating the coefficient of n^3, D = 1;

$$T_n = \frac{1 + 7(n-1) + 6(n-1)(n-2) + 1(n-1)(n-2)(n-3)}{(n-1)!}$$

$$T_n = \frac{1}{(n-1)!} + \frac{7}{(n-2)!} + \frac{6}{(n-1)!} + \frac{1}{(n-4)!}$$

Putting n = 1, 2, 3, … and adding

$$S_\infty = \sum_{n=1}^{\infty} \frac{1}{(n-1)!} + 7\sum_{n=2}^{\infty} \frac{1}{(n-2)!} + 6\sum_{n=3}^{\infty} \frac{1}{(n-3)!}$$

$$+ \sum_{n=4}^{\infty} \frac{1}{(n-4)!} = e + 7e + 6e + e = 15e.$$

3) Sum to infinity the series $\frac{11.14}{10.15.20} + \frac{11.14.17}{10.15.20.25} + \cdots\infty$

Sol: Let $S = \frac{11.14}{10.15.20} + \frac{11.14.17}{10.15.20.25} + \cdots\infty$

$$8S = \frac{8.11.14}{4!} \cdot \frac{1}{5^3} + \frac{8.11.14.17}{5!} \cdot \frac{1}{5^5} + \cdots\infty$$

$$\frac{5.8S}{5} = \frac{5.8.11.14}{4!} \cdot \frac{1}{5^4} + \frac{5.8.11.14.17}{5!} \cdot \frac{1}{5^5} + \cdots \infty$$

$$8S + 1 + \frac{5}{1!} \cdot \frac{1}{5} + \frac{5.8}{2!} \cdot \frac{1}{5^2} + \frac{5.8.11}{3!} \cdot \frac{1}{5^3}$$

$$= 1 + \frac{5}{1!} \cdot \frac{1}{5} + \frac{5.8}{2!} \cdot \frac{1}{5^2} + \frac{5.8.11}{3!} \cdot \frac{1}{5^3} + \frac{5.8.11.14}{4!} \cdot \frac{1}{5^4}$$

$$+ \frac{5.8.11.14.17}{5!} \cdot \frac{1}{5^5} + \cdots \infty$$

Here $p = 5$, $q = 3$, $\dfrac{x}{q} = \dfrac{1}{5}$ $\rightarrow$ $x = \dfrac{3}{5}$

$$8S + 2 + \frac{4}{5} + \frac{44}{75} = (1 - X)^{-p/q}$$

$$= \left[1 - \frac{3}{5} \right]^{\frac{-5}{3}}$$

$$= \left[\frac{2}{5} \right]^{\frac{-5}{3}}$$

$$8S + \frac{254}{75} = \left[\frac{5}{2} \right]^{\frac{5}{3}}$$

$$8S = \left[\frac{5}{2} \right]^{\frac{5}{3}} - \frac{254}{75}$$

$$S = \frac{1}{8} \left[\frac{5}{2} \left[\frac{25}{4} \right]^{\frac{1}{3}} - \frac{254}{75} \right].$$

4)   Sum to infinity the series $\sum_{n=0}^{\infty} \dfrac{5n+1}{(2n+1)!}$.

Sol: Let $S = \sum_{n=0}^{\infty} \dfrac{5n+1}{(2n+1)!}$

General Term $t_n = \dfrac{5n+1}{(2n+1)!}$, $n = 0, 1, 2, ...$

Let $5n + 1 = A + B(2n + 1)$

Put $n = -\dfrac{1}{2}$, we get $-\dfrac{5}{2} + 1 = A$ ➡ $A = -\dfrac{3}{2}$

Put $n = 0$, we get $1 = A + B$ ➡ $B = \dfrac{5}{2}$

Therefore, $5n + 1 = -\dfrac{3}{2} + \dfrac{5}{2}(2n+1)$

$$t_n = \frac{-\dfrac{3}{2} + \dfrac{5}{2}(2n+1)}{(2n+1)!}$$

$$= -\frac{3}{2}\frac{1}{(2n+1)!} + \frac{5}{2}\frac{(2n+1)}{(2n+1)!} = -\frac{3}{2}\frac{1}{(2n+1)!} + \frac{5}{2}\frac{1}{(2n)!}$$

$$t_0 = -\frac{3}{2}\cdot\frac{1}{1!} + \frac{5}{2}$$

$$t_1 = -\frac{3}{2}\cdot\frac{1}{3!} + \frac{5}{2}\cdot\frac{1}{2!}$$

$$t_2 = -\frac{3}{2}\cdot\frac{1}{5!} + \frac{5}{2}\cdot\frac{1}{4!}$$

$$S = t_1 + t_2 + t_3 + ...$$

$$= -\frac{3}{2}\left[\frac{1}{1!}+\frac{1}{3!}+\frac{1}{5!}+\cdots\right]+\frac{5}{2}\left[\frac{1}{1}+\frac{1}{2!}+\frac{1}{4!}+\cdots\right]$$

$$= -\frac{3}{2}\left[\frac{e-e^{-1}}{2}\right]+\frac{5}{2}\left[\frac{e+e^{-1}}{2}\right]$$

$$= \frac{1}{4}\left[-3e+3e^{-1}+5e+5e^{-1}\right]$$

$$= \frac{1}{4}[2e+8e^{-1}]$$

$$= \frac{e}{2}+\frac{2}{e}.$$

5) **Sum to infinity the series** $\left(1+\dfrac{1}{2}\right)+\left(\dfrac{1}{3}+\dfrac{1}{4}\right)\dfrac{1}{9}+\left(\dfrac{1}{5}+\dfrac{1}{6}\right)\dfrac{1}{9^2}+\cdots\infty.$

Sol: $\left(1+\dfrac{1}{2}\right)+\left(\dfrac{1}{3}+\dfrac{1}{4}\right)\dfrac{1}{9}+\left(\dfrac{1}{5}+\dfrac{1}{6}\right)\dfrac{1}{9^2}+\cdots\infty = \left[1+\dfrac{1}{3}\cdot\dfrac{1}{9}+\dfrac{1}{5}\cdot\dfrac{1}{9^2}+\cdots\infty\right]+$

$$\left[\frac{1}{2}+\frac{1}{4}\cdot\frac{1}{9}+\frac{1}{6}\cdot\frac{1}{9^2}+\cdots\infty\right]$$

$$= \left[1+\frac{1}{3}\cdot\frac{1}{3^2}+\frac{1}{5}\cdot\frac{1}{3^4}+\cdots\infty\right]+\frac{1}{2}\left[1+\frac{1}{2.9}+\frac{1}{3.9^2}+\cdots\infty\right]$$

$$= 3\left[\frac{1}{3}+\frac{1}{3}\cdot\frac{1}{3^2}+\frac{1}{5}\cdot\frac{1}{3^5}+\cdots\infty\right]+\frac{9}{2}\left[\frac{1}{9}+\frac{1}{2.9^2}+\frac{1}{3.9^3}+\cdots\infty\right]$$

$$= \frac{3}{2}\log\left[\frac{1+\dfrac{1}{3}}{1-\dfrac{1}{3}}\right]-\frac{9}{2}\log\left[1-\frac{1}{9}\right]$$

$$= \frac{3}{2}\log 2 - \frac{9}{2}\log\left[\frac{8}{9}\right]$$

$$= \frac{3}{2} log2 - \frac{9}{2}[3log2 - 2log3]$$

$$= 9log3 - 12log2$$

6) Show that $log_e (1+\frac{1}{n})^n = 1 - \frac{1}{2(n+1)} - \frac{1}{2.3(n+1)^2} - \frac{1}{3.4(n+1)^3} - \cdots \infty.$

Sol: $RHS = 1 - \frac{1}{2(a+1)} - \frac{1}{2.3(n+1)^2} - \frac{1}{3,4(n+1)^3} - \cdots \infty$

$$= 1 - \left[1 - \frac{1}{2}\right]\frac{1}{n+1} - \left[\frac{1}{2} - \frac{1}{3}\right]$$

$$\frac{1}{(n+1)^2} - \left[\frac{1}{3} - \frac{1}{4}\right]\frac{1}{(n+1)^3} - \cdots \infty$$

$$= \left[1 + \frac{1}{2(n+1)} + \frac{1}{3(n+1)^2} + \cdots \infty\right] -$$

$$\left[\frac{1}{n+1} + \frac{1}{2(n+1)} + \frac{1}{3(n+1)^3} + \cdots \infty\right]$$

$$= (n+1)\left[\frac{1}{n+1} + \frac{1}{2(n+1)^2} + \frac{1}{3(n+1)^3} + \cdots \infty\right] -$$

$$\left[\frac{1}{n+1} + \frac{1}{2(n+1)^2} + \frac{1}{3(n+1)^3} + \cdots \infty\right]$$

$$= (n+1-1)\left[\frac{1}{n+1} + \frac{1}{2(n+1)^2} + \frac{1}{3(n+1)^3} + \cdots \infty\right]$$

$$= -n \, log\left[1 - \frac{n}{n+1}\right]$$

$$= -n \, \log\left[\frac{n}{n+1}\right]$$

$$= n \, \log\left[\frac{n+1}{n}\right] \qquad = n \, \log\left[1+\frac{1}{n}\right] \qquad = \log\left[1+\frac{1}{n}\right]$$

$$= \log\left[1+\frac{1}{n}\right]^{n} \qquad = LHS.$$

7) Prove that $\log\sqrt{12} = 1+\left(\frac{1}{2}+\frac{1}{3}\right)\frac{1}{4}+\left(\frac{1}{4}+\frac{1}{5}\right)\frac{1}{4^2}+\left(\frac{1}{6}+\frac{1}{7}\right)\frac{1}{4^3}+\cdots\infty.$

Sol: Let $S = 1+\left(\frac{1}{2}+\frac{1}{3}\right)\frac{1}{4}+\left(\frac{1}{4}+\frac{1}{5}\right)\frac{1}{4^2}+\left(\frac{1}{6}+\frac{1}{7}\right)\frac{1}{4^3}+\cdots\infty$

$$= \left[1+\frac{1}{3}.\frac{1}{4}+\frac{1}{5}.\frac{1}{4^2}+\frac{1}{7}.\frac{1}{4^3}+L\right]+\left[\frac{1}{2}.\frac{1}{4}+\frac{1}{4}.\frac{1}{4^2}+\frac{1}{6}.\frac{1}{4^3}+\cdots\right]$$

Take $x = \frac{1}{2}.$

Therefore $S = \left[1+\frac{1}{3}.x^2+\frac{1}{5}.x^4+\frac{1}{7}.x^6+\cdots\right]+\left[\frac{1}{2}.x^2+\frac{1}{4}.x^4+\frac{1}{6}.x^6+\cdots\right]$

$$= \frac{1}{x}\left[\frac{x}{1}+\frac{x^3}{3}+\frac{x^5}{5}+\frac{x^7}{7}+\cdots\right]+\frac{1}{2}\left[\frac{x^2}{1}+\frac{x^4}{2}+\frac{x^6}{3}+\cdots\right]$$

$$= \frac{1}{x}\{\frac{1}{2}\log[\frac{1+x}{1-x}]\}+\frac{1}{2}\{\log(1-x^2)\}$$

$$= \frac{1}{2x}\log[\frac{1+x}{1-x}]-\frac{1}{2}\log\left(1-x^2\right)$$

$$= \log\left[\frac{1+\frac{1}{2}}{1-\frac{1}{2}}\right]-\frac{1}{2}\log\left[1-\frac{1}{4}\right]$$

$$= \log 3 - \log \left[\frac{3}{4}\right]^{1/2}$$

$$= \log 3 - \log \sqrt{\frac{3}{4}}$$

$$= \log \left[\frac{3}{\sqrt{\frac{3}{4}}}\right]$$

$$= \log \left[3 \times \sqrt{\frac{4}{3}}\right]$$

$$= \log \sqrt{12}.$$

THEORY OF EQUATIONS

2 MARKS

1) Form the cubic equation two of whose roots are $3, 1+\sqrt{2}$.

Sol: Irrational roots occurs in conjugate pair, then $3, 1+\sqrt{2}, 1-\sqrt{2}$ are the roots.

$$s_1 = 3+1+\sqrt{2}+1-\sqrt{2} = 5$$

$$s_2 = 3(1+\sqrt{2})+3(1-\sqrt{2})+(1+\sqrt{2})(1-\sqrt{2}) = 6+1-2 = 5$$

$$s_3 = 3(1+\sqrt{2})(1-\sqrt{2}) = -3$$

Hence, required equation is

$$x^3 - s_1 x^2 + s_2 x - s_3 = 0 \Rightarrow x^3 - 5x^2 + 5x + 3 = 0.$$

2) Solve the equation $x^3 - 12x^2 + 39x - 28 = 0$ whose roots are in A.P.

Sol: Let $\alpha - \beta,\ \alpha,\ \alpha + \beta$ be the roots.

Then $\alpha - \beta, + \alpha + \alpha + \beta = 12 \Rightarrow \alpha = 4$

4	1	-12	39	-28
	0	4	-32	28
	1	-8	7	0

Therefore, 4 is a root of the given equation. Let us remove this root by synthetic division.

$$(x - 1)(x - 7) = 0$$

$$x = 1, 7$$

The other roots are given by $x^2 - 8x + 7 = 0$

Therefore, the roots of the equation are 1, 4, 7.

3)   Increase by 7 the roots of the equation

$3x^4 + 7x^3 - 15x^2 + x - 2 = 0$

Sol: Given: $3x^4 + 7x^3 - 15x^2 + x - 2 = 0$ ➜ (1)

To increase the roots of (1) by 7

-7	3	7	-15	1	-2
	0	-21	98	-581	4060
	3	-14	83	-580	4058
	0	-21	245	-2296	
	3	-35	328	-2876	
	0	-21	392		
	3	-56	720		
	0	-21			
	3	-77			

Hence the required equation is $3x^4 - 77x^3 + 720x^2 - 2876x + 4058 = 0$.

5) Define Reciprocal Equation.

Sol: A reciprocal equation is an equation in which the reciprocal of any root is also a root. i.e., For a reciprocal equation $f(x) = 0$, if $a_1, a_2, \ldots\ldots, a_n$ are the roots, then their reciprocal is also the roots of the function $f(x) = 0$.

6) Diminish by 2 the roots of the equation $x^4 + x^3 - 3x^2 + 2x - 4 = 0$

Sol: Given: $x^4 + x^3 - 3x^2 + 2x - 4 = 0$ ➡ (1)

To diminish the roots of (1) by 2

2	1	1	-3	2	4
	0	2	6	6	16
	1	3	3	8	12
	0	2	10	26	
	1	5	13	34	
	0	2	14		
	1	7	27		
	0	2			
	1	9			

Hence the required equation is $x^4 + 9x^3 + 27x^2 + 34x + 12 = 0$.

6) One of the roots of the equation $3x^5 - 4x^4 - 42x^3 + 56x^2 + 27x - 36 = 0$ is $\sqrt{2} + \sqrt{5}$. Find the other roots.

Sol: Given that $\sqrt{2} + \sqrt{5}$ is a root.

Therefore, $\sqrt{2} - \sqrt{5}, -\sqrt{2} + \sqrt{5}, -\sqrt{2} - \sqrt{5}$ are also roots.

The real factor corresponding to these roots is

$$(x+\sqrt{2}+\sqrt{5})(x+\sqrt{2}-\sqrt{5})(x-\sqrt{2}+\sqrt{5})(x-\sqrt{2}-\sqrt{5})$$

$$=\left(\left(x+\sqrt{2}\right)^2-5\right)\left(\left(x-\sqrt{2}\right)^2-5\right)$$

$$=(x^2+2+2\sqrt{2}x-5)(x^2+2-2\sqrt{2}x-5)$$

$$=(x^2+2\sqrt{2}x-3)(x^2-2\sqrt{2}x-3)$$

$$=x^4-2\sqrt{2}x^3-3x^2+2\sqrt{2}x^3-8x^2-6\sqrt{2}x-3x^2+6\sqrt{2}x+9$$

$$=x^4-14x^2+9$$

$3x^5-4x^4-42x^3+56x^2+27x-36 = 0$ divided by x^4-14x^2+9 We get $3x-4 = 0$.

➡ $x = \dfrac{4}{3}$

Therefore, the roots of the given equation are

$$\sqrt{2}+\sqrt{5},\sqrt{2}-\sqrt{5},-\sqrt{2}+\sqrt{5},-\sqrt{2}-\sqrt{5},\frac{4}{3}.$$

7) Diminish by 1 the roots of the equation $x^4 - 4x^3 - 7x^2 + 22x + 24 = 0$

Sol: Given: $x^4 - 4x^3 - 7x^2 + 22x + 24 = 0$ ➡   (1)

To diminish the roots of (1) by 1

1	1	-4	-7	22	24
	0	1	-3	-10	12
	1	-3	-10	12	36
	0	1	-2	-12	
	1	-2	-12	0	

0	1	-1
1	-1	-13
0	1	
1	0	

Hence the required equation is $x^4 - 13x^2 + 36 = 0$

8)   Solve $x^4 - 11x^2 + 2x + 12 = 0$, given that $\sqrt{5} - 1$ is a root.

Sol: Given: $x^4 - 11x^2 + 2x + 12 = 0$

Since $\sqrt{5} - 1$, is a root, $-\sqrt{5} - 1$ is also a root.

The real factor corresponding to these roots is

$$X^2 - 2x - 3$$

$$
x^2 + 2x - 4 \overline{\big)\ x^4 + 0\,x^3 - 11x^2 + 2x + 12}
$$

$$x^4 + 2x^3 - 4x^2$$

$$-2x^3 - 7x^2 + 2x$$

$$-2x^3 - 4x^2 + 8x$$

$$-3x^2 - 6x + 12$$

$$-3x^2 - 6x + 12$$

$$0$$

$$(x + 1 - \sqrt{5})(x + 1 + \sqrt{5}) = (x + 1)^2 - 5.$$

$$= X^2 + 2x - 4.$$

Therefore, the other roots are given by $x^2 - 2x - 3 = 0$

$(x - 3)(x + 1) = 0$

$x = 3, -1$

Therefore, the roots of the equation are $\sqrt{5} - 1, -\sqrt{5} - 1, 3, -1$.

9) Find the polynomial equation of the least degree having -1, 1, 2 and 3 as roots.

Sol: The polynomial equation of the least degree having -1, 1, 2 and 3 as roots must be a multiple of $(x + 1)(x - 1)(x - 2)(x - 3)$

$= (x^2 - 1)(x^2 - 5x + 6)$

$= x^4 - 5x^3 + 5x^2 + 5x - 6.$

5 MARKS

1) Solve $27x^3 + 42x^2 - 28x - 8 = 0$, given that its roots are in G.P.

Sol: Given: $27x^3 + 42x^2 - 28x - 8 = 0$ ➜   (1)

The roots of (1) are in GP

Let $\dfrac{a}{r}, a, ar$ be the roots.

Sum of the roots: $\dfrac{a}{r} + a + ar = \dfrac{-42}{27}$

$a\left[\dfrac{r^2 + r + 1}{r}\right] = \dfrac{-14}{9}$ ➜   (2)

Product of the roots: $\left(\dfrac{a}{r}\right)a(ar) = -\left(\dfrac{-8}{27}\right)$

$a^3 = \dfrac{8}{27}$

$a = \dfrac{2}{3}$ ➜   (3)

Substituting (3) in (2), we get $\dfrac{2}{3}\left[\dfrac{r^2+r+1}{r}\right]=\dfrac{-14}{9}$

$$\left[\dfrac{r^2+r+1}{r}\right]=\dfrac{-14}{9}\cdot\dfrac{3}{2}$$

$$\left[\dfrac{r^2+r+1}{r}\right]=\dfrac{-7}{3}$$

$$3r^2+3r+3=-7r$$

$$3r^2+10r+3=0$$

$$r=-3,\dfrac{-1}{3}$$

If $r=-3, a=\dfrac{2}{3}$, then the roots are $\dfrac{-2}{6},\dfrac{2}{3},-2$.

If $r=\dfrac{-1}{3}, a=\dfrac{2}{3}$, then the roots are $-2,\dfrac{2}{3},\dfrac{-2}{6}$.

2) Diminish the roots of the equation $x^4 - 4x^3 - 7x^2 + 22x + 24 = 0$ by 1 and hence solve the equation.

Sol: Given: $x^4 - 4x^3 - 7x^2 + 22x + 24 = 0$ ➔ (1)

To diminish the roots of (1) by 1

1	1	-4	-7	22	24
	0	1	-3	-10	12
	1	-3	-10	12	36
	0	1	-2	-12	
	1	-2	-12	0	

0	1	-1
1	-1	-13
0	1	
1	0	

Hence the required equation is $x^4 - 13x^2 + 36 = 0$

$$(x^2 - 4)(x^2 - 9) = 0$$

$$x^2 = 4, 9$$

$$x = -2, 2, -3, 3$$

Therefore, the roots of the required equation are $-2, 2, -3, 3$

Therefore, the roots of the given equation are $-2, -1, 3, 4$.

3) Remove the second term of the equation $x^3 - 6x^2 + 11x - 6 = 0$.

Sol: Let us diminish the roots by h so that the second term is removed.

Let $y = x - h$. Therefore, $x = y + h$.

The given equation becomes $(y + h)^3 - 6(y + h)^2 + 11(y + h) - 6 = 0$

$$y^3 + y^2(3h - 6) + y(3h^2 - 12h + 11) + h^3 - 6h^2 + 11h - 6 = 0$$

If the second term is to be absent, then $3h - 6 = 0 \rightarrow h = 2$

Diminish the roots by 2

2	1	-6	11	-6
	0	2	-8	6
	1	-4	3	0
	0	2	-4	

1	-2	-1
0	2	
1	0	

Therefore, the required equation is $x^3 - x = 0$.

4) If α is a root of the equation $x^3 + x^2 - 2x - 1 = 0$, then show that $\alpha^2 - 2$ is also a root.

Sol: Given: $x^3 + x^2 - 2x - 1 = 0$ ➜ (1)

Let us form the equation whose roots are $\alpha^2 - 2$, $\beta^2 - 2$, $^2 - 2$.

Let $y = \alpha^2 - 2$ ➜ $y = x^2 - 2$ ➜ $x^2 = y + 2$.

The equation becomes $x(y + 2) + y + 2 - 2x - 1 = 0$

$$xy = - (y + 1)$$

$$x^2y^2 = (y + 1)^2$$

$$(y + 2)y^2 = (y + 1)^2$$

$$y^3 + y^2 - 2y - 1 = 0 \quad ➜ \quad (2)$$

Comparing (1) & (2), we note that the roots of the two equations are the same.

Therefore, $\alpha^2 - 2$ is also a root of the given equation.

5) Show that the roots of the equation $x^3 + px^2 + qx + r = 0$ are in A.P.

Then $2p^3 - 9pq + 27r = 0$.

Sol: Given: $x^3 + px^2 + qx + r = 0$ ➜ (1)

Let the roots of (1) be $a - d$, a, $a + d$.

Sum of the roots: $a - d + a + a + d = - p$

$$3a = -p$$

$$\rightarrow \quad a = \frac{-p}{3}$$

But 'a' is the roots of the given equation.

Therefore, $a^3 + pa^2 + qa + r = 0$

$$\rightarrow \quad \left[\frac{-p}{3}\right]^3 + p\left[\frac{-p}{3}\right]^2 + q\left[\frac{-p}{3}\right] + r = 0$$

$$\frac{-p^3}{27} + \frac{p^3}{9} - \frac{pq}{3} + r = 0$$

$$-p^3 + 3p^2 - 9pq + 27r = 0$$

$$\rightarrow 2p^3 - 9pq + 27r = 0.$$

6) Solve the equation $4x^4 - 20x^3 + 33x^2 - 20x + 4 = 0$.

Sol: Given: $4x^4 - 20x^3 + 33x^2 - 20x + 4 = 0$

The given equation is of first class and even degree.

Dividing by x^2 and grouping, we get

$$4\left[x^2 + \frac{1}{x^2}\right] - 20\left[x + \frac{1}{x}\right] + 33 = 0 \quad \rightarrow \qquad (1)$$

Put $x + \dfrac{1}{x} = y$ and $x^2 + \dfrac{1}{x^2} = y^2 - 2$ in $\qquad (1)$

$$\rightarrow 4(y^2 - 2) - 20y + 33 = 0$$

$$4y^2 - 8 - 20y + 33 = 0$$

$$4y^2 - 20y + 25 = 0$$

$$(2y - 5)^2 = 0$$

$$y = \frac{5}{2}, \frac{5}{2}$$

$$x + \frac{1}{x} = y$$

$$\rightarrow \quad x + \frac{1}{x} = \frac{5}{2}$$

$$2x^2 - 5x + 2 = 0$$

$$(2x - 1)(x - 2) = 0$$

$$\rightarrow \quad x = 2, \frac{1}{2}$$

Therefore, the roots of the given equation are $2, \frac{1}{2}, 2, \frac{1}{2}$.

7) Solve the equation $2x^3 - x^2 - 22x - 24 = 0$ given that two of its roots are in the ratio 3:4

Sol: Given: $2x^3 - x^2 - 22x - 24 = 0$

Let the roots of the equation be $3\alpha, 4\alpha, \beta$

Then $7\alpha + \beta = \frac{1}{2} \rightarrow$ $\qquad\qquad\qquad\qquad$ (1)

$12\alpha^2 + 7\alpha\beta = -11 \rightarrow$ $\qquad\qquad\qquad\qquad$ (2)

$12\alpha^2\beta = 12 \rightarrow \alpha^2\beta = 1 \rightarrow$ $\qquad\qquad\qquad$ (3)

From (1), $\beta = \frac{1}{2} - 7\alpha$

$$12\alpha^2 + 7\alpha\left[\frac{1}{2} - 7\alpha\right] = -11$$

$$12\alpha^2 + \frac{7\alpha}{2} - 49\alpha^2 = -11$$

$$24\alpha^2 + 7\alpha - 98\alpha^2 = -22$$

$$74\alpha^2 + 7\alpha - 22 = 0$$

$$(2\alpha + 1)(37\alpha - 22) = 0$$

$$\alpha = -\frac{1}{2}(or)\frac{22}{37}$$

When $\alpha = -\frac{1}{2}, \beta = 4$

Substituting these values in (3), we get $\frac{1}{4} \times 4 = 1$ which is true.

Therefore, the roots are $-\frac{3}{2}, -2, 4.$

$\left(\alpha = \frac{22}{37} \text{ does not satisfy the given equation.}\right)$

8) Find the real roots of the equation $x^3 - 3x + 1 = 0$ using Newton – Raphson method.

Sol: Given: $x^3 - 3x + 1 = 0$

Let $f(x) = x^3 - 3x + 1 = 0$

$$f'(x) = 3x^2 - 3$$

$$f(0) = 1 > 0$$

$$f(1) = -1 < 0$$

Therefore, the root lies between 0 & 1. So $x_0 = 0.5$

We know that $x_{n+1} = x_n - \dfrac{f(x_n)}{f'(x_n)}$

I – Iteration: n = 0, $x_0 = 0.5$

$$x_1 = x_0 - \frac{f(x_0)}{f'(x_0)} = 0.5 - \frac{f(0.5)}{f'(0.5)} = 0.5 - \frac{0.375}{2.250} = 0.333$$

II – Iteration: n = 1, $x_1 = 0.333$

$$x_2 = x_1 - \frac{f(x_1)}{f'(x_1)} = 0.333 - \frac{f(0.333)}{f'(0.333)} = 0.333 + \frac{0.038}{2.667} = 0.347$$

III – Iteration: n = 2, x_2 = 0.347

$$x_3 = x_2 - \frac{f(x_2)}{f'(x_2)} = 0.347 - \frac{f(0.347)}{f'(0.347)} = 0.347 + \frac{0.001}{2.639} = 0.347$$

Therefore, the real root of the equation is 0.347 approximately.

9) Solve the equation $x^4 - 4x^2 + 8x + 35 = 0$ given that the roots is $2 + i\sqrt{3}$.

Sol: Given: $x^4 - 4x^2 + 8x + 35 = 0$

Since $2 + i\sqrt{3}$ is a root, $2 - i\sqrt{3}$ is also a root.

The real factor corresponding to these roots is

$$(x - 2 - i\sqrt{3})(x - 2 + i\sqrt{3}) = (x - 2)^2 + 3 = x^2 - 4x + 7$$

Therefore, the other roots are given by $x^2 + 4x + 5 = 0$

$$x = \frac{-4 \pm \sqrt{16 - 20}}{2} = \frac{-4 \pm 2i}{2} = -2 \pm i$$

$$X^2 + 4x + 5$$

$$x^2 - 4x + 7 \quad x^4 + 0\,x^3 - 4x^2 + 8x + 35$$

$$x^4 - 4x^3 + 7x^2$$

$$4x^3 - 11x^2 + 8x$$

$$4x^3 - 16x^2 + 28x$$

$$5x^2 - 20x + 35$$

$$5x^2 - 20x + 35$$

$$0$$

Therefore, the roots of the given equation are $2 \pm i\sqrt{3}, -2 \pm i$

10) Find the roots of the equation $x^3 + 6x - 2 = 0$ using Newton – Raphson method.

Sol: Given: $x^3 + 6x - 2 = 0$

Let $f(x) = x^3 + 6x - 2 = 0$

$f(x) = 3x^2 + 6$

$f(0) = -2 < 0$

$f(1) = 5 > 0$

Therefore, the root lies between 0 & 1. So $x_0 = 0.5$

We know that $x_{n+1} = x_n - \dfrac{f(x_n)}{f^|(x_n)}$

I – Iteration: n = 0, $x_0 = 0.5$

$$x_1 = x_0 - \frac{f(x_0)}{f^|(x_0)} = 0.5 - \frac{f(0.5)}{f^|(0.5)} = 0.5 - \frac{1.125}{6.750} = 0.333$$

II – Iteration: n = 1, $x_1 = 0.333$

$$x_2 = x_1 - \frac{f(x_1)}{f^|(x_1)} = 0.333 - \frac{f(0.333)}{f^|(0.333)} = 0.333 - \frac{0.035}{6.333} = 0.327$$

III – Iteration: n = 2, $x_2 = 0.327$

$$x_3 = x_2 - \frac{f(x_2)}{f^|(x_2)} = 0.327 - \frac{f(0.327)}{f^|(0.327)} = 0.327 + \frac{0.003}{6.327} = 0.327$$

Therefore, the real root of the equation is 0.327 approximately.

11) Solve $x^4 - 5x^3 + 4x^2 + 8x - 8 = 0$ given that one of its roots is $\sqrt{5} + 1$.

Sol: Given: $x^4 - 5x^3 + 4x^2 + 8x - 8 = 0$

Since $\sqrt{5} + 1$ is a root, $-\sqrt{5} + 1$ is also a root.

Let the other roots be α & β.

Sum of the roots: $\sqrt{5} + 1 - \sqrt{5} + 1 + \alpha + \beta = 5$ ➜ $\alpha + \beta = 3$

Product of the roots: $(\sqrt{5} + 1)(-\sqrt{5} + 1)\,\alpha\beta = -8$ ➜ $\alpha\beta = 2$

Therefore, α & β are the roots of $x^2 - x(\alpha + \beta) + \alpha\beta = 0$

$x^2 - 3x + 2 = 0$

$(x - 1)(x - 2) = 0$ ➜ $x = 1, 2$

Therefore, the roots of the given equation are $\sqrt{5} + 1, -\sqrt{5} + 1$ 1, 1, 2.

13) If the roots of $x^3 + px^2 + qx + \lambda = 0$ are in G.P., then show that $rp^3 = q^3$.

Sol: Given: $x^3 + px^2 + qx + \lambda = 0$ ➜ $\qquad\qquad\qquad$ (1)

Let α, β, δ be the roots of $\qquad\qquad\qquad\qquad\qquad\qquad$ (1)

If the roots are in G.P., then $\beta^2 = \alpha\delta$ ➜ $\qquad\qquad\qquad$ (2)

Product of the roots: $\alpha\beta\delta = -r$ ➜ $\beta^3 = -r$ (by (2)) ➜ $\qquad$ (3)

Since, β is a root of (1), we get $\beta^3 + p\beta^2 + q\beta + r = 0$

$-r + p\beta^2 + q\beta + r = 0$

$\beta\,(p\beta + q) = 0$

But $\beta \neq 0$

Therefore, $p\beta + q = 0$ ➜ $\beta = -\dfrac{q}{p}$

$\beta^3 = -\dfrac{q^3}{p^3}$ ➜ $\qquad\qquad\qquad\qquad\qquad\qquad\qquad\qquad$ (4)

Comparing (3) and (4), we get $rp^3 = q^3$, which is the required condition.

13) Find the equation whose roots are the roots of the equation $4x^4 + 32x^3 + 83x^2 + 76x + 21 = 0$ increased by 2 and hence solve the given equation.

Sol: Given: $4x^4 + 32x^3 + 83x^2 + 76x + 21 = 0$

To increase the roots by 2

-2	4	32	83	76	21
	0	-8	-48	-70	-12
	4	24	35	6	9
	0	-8	-32	-6	
	4	16	3	0	
	0	-8	-16		
	4	8	-13		
	0	-8			
	4	0			

Therefore, the required equation is $4x^4 - 13x^2 + 9 = 0$

$$4x^4 - 4x^2 - 9x^2 + 9 = 0$$

$$4x^2(x^2 - 1) - 9(x^2 - 1) = 0$$

$$(4x^2 - 9)(x^2 - 1) = 0 \implies x = 1, -1, \frac{3}{2}, -\frac{3}{2}$$

Therefore, the roots of the given equation are $-3, -1, -\frac{7}{2}, -\frac{1}{2}$.

14) Solve $x^5 - 6x^4 + 7x^3 + 7x^2 - 6x + 1 = 0$.

Sol: Given: $x^5 - 6x^4 + 7x^3 + 7x^2 - 6x + 1 = 0$

This is a Reciprocal Equation of First Class and Odd Degree.

Therefore, $x = -1$ is a root.

-1	1	-6	7	7	-6	1
	0	-1	7	-14	7	-1
	1	-7	14	-7	1	0

Therefore, the other roots are given by $x^4 - 7x^3 + 14x^2 - 7x + 1 = 0$

Dividing by x^2, $\left[x^2 + \dfrac{1}{x^2}\right] - 7\left[x + \dfrac{1}{x}\right] + 14 = 0$ ➜ (1)

Put $x + \dfrac{1}{x} = y$ and $x^2 + \dfrac{1}{x^2} = y^2 - 2$ in (1)

$y^2 - 2 - 7y + 14 = 0$

$y^2 - 7y + 12 = 0$ ➜ $y = 3$ or 4.

$$x + \frac{1}{x} = 3 \qquad\qquad x + \frac{1}{x} = 4$$

$$x^2 - 3x + 1 = 0 \qquad\qquad x^2 - 4x + 1 = 0$$

$$x = \frac{3 \pm \sqrt{5}}{2} \qquad\qquad x = 2 \pm \sqrt{3}$$

Therefore, the roots of the given equation are $\dfrac{3 \pm \sqrt{5}}{2}, 2 \pm \sqrt{3}$.

10 MARKS

1) Find the real root of the equation $x^3 - 2x^2 - 3x - 4 = 0$ by Newton – Raphson method.

Sol: Given: $x^3 - 2x^2 - 3x - 4 = 0$

$f(x) = x^3 - 2x^2 - 3x - 4 = 0$

$f^{l}(x) = 3x^2 - 4x - 3$

$$f(0) = -4 < 0$$

$$f(1) = -8 < 0$$

$$f(2) = -10 < 0$$

$$f(3) = -4 < 0$$

$$f(4) = 16 > 0$$

Therefore, the root lies between 3 and 4. So $x_0 = 3.5$

We know that $x_{n+1} = x_n - \dfrac{f(x_n)}{f^{|}(x_n)}$

I – Iteration: n = 0, $x_0 = 3.5$

$$x_1 = x_0 - \frac{f(x_0)}{f^{|}(x_0)}$$

$$= 3.5 - \frac{f(3.5)}{f^{|}(3.5)}$$

$$= 3.5 - \frac{3.875}{19.750} = 3.304$$

II – Iteration: n = 1, $x_1 = 3.304$

$$x_2 = x_1 - \frac{f(x_1)}{f^{|}(x_1)}$$

$$= 3.304 - \frac{f(3.304)}{f^{|}(3.304)}$$

$$= 3.304 - \frac{0.323}{16.533} = 3.284$$

III – Iteration: n = 2, $x_2 = 3.284$

$$x_3 = x_2 - \frac{f(x_2)}{f^{|}(x_2)}$$

$$= 3.284 - \frac{f(3.284)}{f^{|}(3.284)}$$

$$= 3.284 + \frac{0.005}{16.218} = 3.284$$

Therefore, the real root of the equation is 3.284 approximately.

2) Evaluate $\sqrt{12}$ to four decimal places by Newton's method.

Sol: Let $x = \sqrt{12}$ ➜ $x^2 - 12 = 0$

$f(x) = x^2 - 12$ ➜ $f^{|}(x) = 2x$

$f(0) = -12 < 0$

$f(1) = -11 < 0$

$f(2) = -8 < 0$

$f(3) = -3 < 0$

$f(4) = 4 > 0$

Therefore, the roots lies between 3 and 4. Let $x_0 = 3$

We know that $x_{n+1} = x_n - \frac{f(x_n)}{f^{|}(x_n)}$

I – Iteration: $x_1 = x_0 - \frac{f(x_0)}{f^{|}(x_0)}$

$$= 3 - \frac{f(3)}{f^{|}(3)}$$

$$= 3 - \frac{(-3)}{6} = 3.5$$

II – Iteration: $x_2 = x_1 - \dfrac{f(x_1)}{f^|(x_1)}$

$$= 3.5 - \dfrac{f(3.5)}{f^|(3.5)}$$

$$= 3.5 - \dfrac{5.25}{7} = 3.464$$

III – Iteration: $x_3 = x_2 - \dfrac{f(x_2)}{f^|(x_2)}$

$$= 3.464 - \dfrac{f(3.464)}{f^|(3.464)}$$

$$= 3.464 - \dfrac{0.000}{6.928} = 3.464$$

Therefore, $\sqrt{12} = 3.464$.

3)   Solve $6x^5 - x^4 - 43x^3 + 43x^2 + x - 6 = 0$.

Sol: Given: $6x^5 - x^4 - 43x^3 + 43x^2 + x - 6 = 0$

This is a Reciprocal Equation of odd Degree with unlike signs.

Therefore, 1 is a root of this equation.

1	6	-1	-43	43	1	-6
	0	6	5	-38	5	6
	6	5	-38	5	6	0

Therefore, the other roots are given by $6x^4 + 5x^3 - 38x^2 + 5x + 6 = 0$

Dividing by x^2, $6\left[x^2 + \dfrac{1}{x^2}\right] - 5\left[x + \dfrac{1}{x}\right] - 38 = 0$ ➡   (1)

Put $x + \dfrac{1}{x} = y$ and $x^2 + \dfrac{1}{x^2} = y^2 - 2$ in   (1)

$$6(y^2 - 2) + 5y - 38 = 0$$

$$6y^2 + 5y - 50 = 0 \quad \rightarrow \quad y = -\frac{10}{3} \, or \, \frac{5}{2}$$

$$x + \frac{1}{x} = -\frac{10}{3} \qquad\qquad x + \frac{1}{x} = \frac{5}{2}$$

$$3x^2 + 10x + 3 = 0 \qquad\qquad 2x^2 - 5x + 2 = 0$$

$$x = -3, -\frac{1}{3} \qquad x = -2, \frac{1}{2}$$

Therefore, the roots of the given equation are $1, -3, -\frac{1}{3}, 2, \frac{1}{2}$.

4) Find by Newton's method an approximate value of the positive root of the equation

$$x^3 - 2x - 5 = 0.$$

Sol: Given: $x^3 - 2x - 5 = 0$

Let $f(x) = x^3 - 2x - 5$

$f^|(x) = 3x^2 - 2$

$f(0) = -5 < 0$

$f(1) = -6 < 0$

$f(2) = -1 < 0$

$f(3) = 16 > 0$

Therefore, the root lies between 2 and 3. So $x_0 = 2.5$

We know that $x_{n+1} = x_n - \dfrac{f(x_n)}{f^|(x_n)}$

I – Iteration: n = 0, $x_0 = 2.5$

$$x_1 = x_0 - \frac{f(x_0)}{f^|(x_0)}$$

$$= 2.5 - \frac{f(2.5)}{f^{|}(2.5)}$$

$$= 2.5 - \frac{5.625}{16.750} = 2.164$$

II – Iteration: n = 1, x_1 = 2.164

$$\mathbf{x}_2 = \mathbf{x}_1 - \frac{f(x_1)}{f^{|}(x_1)}$$

$$= 2.164 - \frac{f(2.164)}{f^{|}(2.164)}$$

$$= 2.164 - \frac{0.806}{12.049} = 2.097$$

III – Iteration: n = 2, x_2 = 2.097

$$\mathbf{x}_3 = \mathbf{x}_2 - \frac{f(x_2)}{f^{|}(x_2)}$$

$$= 2.097 - \frac{f(2.097)}{f^{|}(2.097)}$$

IV – Iteration: n = 2, x_2 = 2.095

$$\mathbf{x}_3 = \mathbf{x}_2 - \frac{f(x_2)}{f^{|}(x_2)}$$

$$= 2.095 - \frac{f(2.095)}{f^{|}(2.095)}$$

$$= 2.095 - \frac{0.005}{11.167} = 2.095$$

Therefore, the real root of the equation is 2.095 approximately.

5) Solve $3x^6 + x^5 - 27x^4 + 27x^2 - x - 3 = 0$.

Sol: Given: $3x^6 + x^5 - 27x^4 + 27x^2 - x - 3 = 0$

This is a reciprocal equation of even degree with the middle term is missing.

Therefore, $1, -1$ are the roots.

1	3	1	-27	0	27	-1	-3
	0	3	4	-23	-23	4	3
	3	4	-23	-23	4	3	0
	0	-3	-1	24	-1	-3	
	3	1	-24	1	3	0	

The other roots are given by $3x^4 + x^3 - 24x^2 + x + 3 = 0$.

Dividing by x^2, $\quad 3\left[x^2 + \dfrac{1}{x^2}\right] + \left[x + \dfrac{1}{x}\right] - 24 = 0 \hspace{3cm} (1)$

Put $\quad x + \dfrac{1}{x} = y$ and $x^2 + \dfrac{1}{x^2} = y^2 - 2$ in

$3(y^2 - 2) + y - 24 = 0$

$3y^2 + y - 30 = 0$

$(y - 3)(3y + 10) = 0 \quad \Rightarrow \quad y = 3, -\dfrac{10}{3}$

$$x + \dfrac{1}{x} = -\dfrac{10}{3} \hspace{3cm} x + \dfrac{1}{x} = 3$$

$$3x^2 + 10x + 3 = 0 \hspace{3cm} x^2 - 3x + 1 = 0$$

$$x = -3, -\frac{1}{3} \qquad x = \frac{3 \pm \sqrt{5}}{2}$$

Therefore, the roots of the given equation are $1, -1, -3, -\frac{1}{3}, \frac{3 \pm \sqrt{5}}{2}$.

6)   If the sum of the two roots of the equation $x^4 + px^3 + qx^2 + rx + s = 0$ equals the sum of the other two, then prove that $p^3 + 8r = 4pq$.

Sol: Let $\alpha, \beta, \gamma, \delta$ be the roots such that $\alpha + \beta = \gamma + \delta$ ➡ (*)

Then $x^4 + px^3 + qx^2 + rx + s = (x - \alpha)(x - \beta)(x - \gamma)(x - \delta)$

$$= [x^2 - (\alpha + \beta)x + \alpha\beta][x^2 - (\gamma + \delta)x + \gamma\delta]$$

Equating the coefficient of x^3, $\alpha + \beta + \gamma + \delta = -p$ ➡   (1)

Equating the coefficient of x^2, $(\alpha+\beta)(\gamma+\delta)+\alpha\beta + \gamma\delta = q$ ➡   (2)

Equating the coefficient of x, $-(\alpha\beta) - (\gamma+\delta) - \gamma\delta(\alpha + \beta) = r$

    ➡ $(\alpha\beta) + (\gamma+\delta) + \gamma\delta(\alpha + \beta) = -r$ ➡   (3)

Equating the constant term $\alpha\beta\gamma\delta, = s$ ➡   (4)

Using (*) in (1), we get $2(\alpha + \beta) = -p$ or $\alpha + \beta = -\frac{p}{2}$.

Therefore, $\gamma + \delta = -\frac{p}{2}$

From (2), $\alpha\beta + \gamma\delta = q - \frac{p^2}{4}$ ➡   (5)

Using (*) in (3), we get $(\alpha\beta + \gamma\delta)\left[-\frac{p}{2}\right] = -r$

Therefore, $\alpha\beta + \gamma\delta = \frac{2r}{p}$ ➡   (6)

From (5) and (6), we get $q - \frac{p^2}{4} = \frac{2r}{p}$

    ➡ $4pq - p^3 = 8r$ ➡ $p^3 + 8r = 4pq$.

UNIT – 3

MATRICES

2 MARKS

1) If $A = \begin{bmatrix} \cos\theta & \sin\theta \\ -\sin\theta & \cos\theta \end{bmatrix}$ then show that A is orthogonal.

Sol: Given: $A = \begin{bmatrix} \cos\theta & \sin\theta \\ -\sin\theta & \cos\theta \end{bmatrix}$

Then $A^T = \begin{bmatrix} \cos\theta & -\sin\theta \\ \sin\theta & \cos\theta \end{bmatrix}$

Now $AA^T = \begin{bmatrix} \cos\theta & \sin\theta \\ -\sin\theta & \cos\theta \end{bmatrix} \begin{bmatrix} \cos\theta & -\sin\theta \\ \sin\theta & \cos\theta \end{bmatrix}$

$= \begin{bmatrix} \cos^2\theta + \sin^2\theta & -\cos\theta\sin\theta + \sin\theta\cos\theta \\ -\sin\theta\cos\theta + \cos\theta\sin\theta & \sin^2\theta + \cos^2\theta \end{bmatrix}$

$= \begin{bmatrix} 1 & 0 \\ 0 & 1 \end{bmatrix} = I$

Therefore, A is orthogonal.

2) State Cayley – Hamilton theorem.

Sol: Every square matrix satisfies its own characteristic equation.

3) Define Scalar matrix and give an example.

Sol: In a diagonal matrix if all the elements are equal, then the matrix is called a scalar matrix.

$$\text{Eg}: \begin{bmatrix} 2 & 0 & 0 \\ 0 & 2 & 0 \\ 0 & 0 & 2 \end{bmatrix}$$

4) If A and B are symmetric, then show that A+B is symmetric.

Sol: Given: A and B are symmetric.

Therefore $A = A^T$ and $B = B^T$

$$(A + B) = A^T + B^T = (A + B)^T$$

➔ A + B is symmetric.

5) Write the transposed conjugate of the matrix $\begin{bmatrix} 2+i & 7+4i & 6-3i \\ 3-2i & 4-3i & 2+3i \\ 6 & 7+6i & 7 \end{bmatrix}$..

Sol: Let $A = \begin{bmatrix} 2+i & 7+4i & 6-3i \\ 3-2i & 4-3i & 2+3i \\ 6 & 7+6i & 7 \end{bmatrix}$.

Conjugate of $A, (\overline{A}) = \begin{bmatrix} 2-i & 7-4i & 6+3i \\ 3+2i & 4+3i & 2-3i \\ 6 & 7-6i & 7 \end{bmatrix}$

Transposed Conjugate, $(\overline{A})^T = \begin{bmatrix} 2-i & 3-2i & 6 \\ 7+4i & 4+3i & 7-6i \\ 6+3i & 2-3i & 7 \end{bmatrix}$.

6) Find the rank of the matrix $A = \begin{bmatrix} 1 & 2 & 3 & 4 \\ 2 & 4 & 6 & 8 \\ -1 & -2 & -3 & -4 \end{bmatrix}$.

Sol: Given: $A = \begin{bmatrix} 1 & 2 & 3 & 4 \\ 2 & 4 & 6 & 8 \\ -1 & -2 & -3 & -4 \end{bmatrix} (R_2 = R_2 - 2R_1)$

$$\begin{bmatrix} 1 & 2 & 3 & 4 \\ 0 & 0 & 0 & 0 \\ -1 & -2 & -3 & -4 \end{bmatrix}$$

$$(R_3 = R_3 + R_1)\begin{bmatrix} 1 & 2 & 3 & 4 \\ 0 & 0 & 0 & 0 \\ 0 & 0 & 0 & 0 \end{bmatrix}$$

Therefore, Rank of A = 1.

7) Define Hermitian matrix.

Sol: A square matrix A = $[a_{ij}]$ is said to be Hermitian if the $(I,j)^{th}$ element of A is equal to the conjugate complex of the $(j,i)^{th}$ element of A.

8) Find the eigen values of A, given that $A = \begin{bmatrix} 1 & 2 & 3 \\ 0 & 2 & -7 \\ 0 & 0 & 3 \end{bmatrix}$.

Sol: Given: $A = \begin{bmatrix} 1 & 2 & 3 \\ 0 & 2 & -7 \\ 0 & 0 & 3 \end{bmatrix}$

The characteristic equation is $|A-\lambda I| = 0$

$$\begin{vmatrix} 1-\lambda & 2 & 3 \\ 0 & 2-\lambda & -7 \\ 0 & 0 & 3-\lambda \end{vmatrix} = 0$$

$(1 - \lambda)(2 - \lambda)(3 - \lambda) = 0$

$\lambda = 1,2,3.$

Therefore, the eigen values are 1, 2, 3.

9) If A and B are unitary matrices, then prove that AB and BA are also unitary matrices.

Sol: Since A and B are unitary matrices, we have

$AA^* = A^*A = I$ and $BB^* = B^*B = I$

Now $(AB)(AB)^* = AB(B^*A^*)$

$\quad = A(BB^*)A^*$

$$= (AI)A^* \quad = AA^* \quad = I$$

Similarly $(BA)(BA)^* = BA(A^*B^*)$

$$= B(AA^*)B^*$$

$$= (BI)B^* \quad = BB^* \quad = I.$$

10) Define a Skew – Symmetric matrix.

Sol: A square matrix $A = (a_{ij})$ is called skew – symmetric if $(a_{ij}) = -(a_{ij})$.

11) Prove that the matrix $\begin{bmatrix} \dfrac{1}{\sqrt{2}} & \dfrac{-1}{\sqrt{2}} \\ \dfrac{1}{\sqrt{2}} & \dfrac{1}{\sqrt{2}} \end{bmatrix}$ is orthogonal.

Sol: Let $A = \begin{bmatrix} \dfrac{1}{\sqrt{2}} & \dfrac{-1}{\sqrt{2}} \\ \dfrac{1}{\sqrt{2}} & \dfrac{1}{\sqrt{2}} \end{bmatrix}$ $A^| = \begin{bmatrix} \dfrac{1}{\sqrt{2}} & \dfrac{1}{\sqrt{2}} \\ \dfrac{-1}{\sqrt{2}} & \dfrac{1}{\sqrt{2}} \end{bmatrix}$

$$AA^| = \begin{bmatrix} \dfrac{1}{\sqrt{2}} & \dfrac{-1}{\sqrt{2}} \\ \dfrac{1}{\sqrt{2}} & \dfrac{1}{\sqrt{2}} \end{bmatrix} \begin{bmatrix} \dfrac{1}{\sqrt{2}} & \dfrac{1}{\sqrt{2}} \\ \dfrac{-1}{\sqrt{2}} & \dfrac{1}{\sqrt{2}} \end{bmatrix} = \begin{bmatrix} 1 & 0 \\ 0 & 1 \end{bmatrix} = I$$

Therefore, A is orthogonal.

12) Define Unitary matrix.

Sol: A square matrix 'A' is said to be unitary if $A(\bar{A})^T = I$.

13) Find the rank of $\begin{bmatrix} -1 & 0 & 2 & 1 \\ 0 & 1 & 1 & -1 \\ 2 & 0 & -4 & -2 \end{bmatrix}$.

Sol: Let $A = \begin{bmatrix} -1 & 0 & 2 & 1 \\ 0 & 1 & 1 & -1 \\ 2 & 0 & -4 & -2 \end{bmatrix} (R_3 = R_3 + 2R_1)$

$$\begin{bmatrix} -1 & 0 & 2 & 1 \\ 0 & 1 & 1 & -1 \\ 0 & 0 & 0 & 0 \end{bmatrix}$$

Therefore, Rank of A = 2.

14) Express $\begin{bmatrix} 2 & 4 & 8 \\ 6 & 2 & 8 \\ 2 & 2 & 2 \end{bmatrix}$ as the sum of a symmetric and a skew – symmetric matrix.

Sol: Let $A = \begin{bmatrix} 2 & 4 & 8 \\ 6 & 2 & 8 \\ 2 & 2 & 2 \end{bmatrix}$ and $A^T = \begin{bmatrix} 2 & 6 & 2 \\ 4 & 2 & 2 \\ 8 & 8 & 2 \end{bmatrix}$

Symmetric : $A = A^T$

Skew – Symmetric : $A = -A^T$.

5 MARKS

1) Find the rank of $\begin{bmatrix} 2 & -1 & 3 & 4 \\ 0 & 3 & 4 & 1 \\ 2 & 3 & 7 & 5 \\ 2 & 5 & 11 & 6 \end{bmatrix}$

Sol: Let $A = \begin{bmatrix} 2 & -1 & 3 & 4 \\ 0 & 3 & 4 & 1 \\ 2 & 3 & 7 & 5 \\ 2 & 5 & 11 & 6 \end{bmatrix}$ $(R_3 = R_3 - (R_1 + R_2))$

$$\begin{bmatrix} 2 & -1 & 3 & 4 \\ 0 & 3 & 4 & 1 \\ 0 & 1 & 0 & 0 \\ 2 & 5 & 11 & 6 \end{bmatrix}$$

$$(C_3 = C_3 - (C_2 + C_4)) \begin{bmatrix} 2 & -1 & 0 & 4 \\ 0 & 3 & 0 & 1 \\ 0 & 1 & -1 & 0 \\ 2 & 5 & 0 & 6 \end{bmatrix}$$

Therefore, Rank of A = 4.

2)   Using Cayley – Hamilton Theorem, find A^{-1} if $A = \begin{bmatrix} 1 & -1 & 2 \\ -2 & 1 & 3 \\ 3 & 2 & -3 \end{bmatrix}$.

Sol: Given: $A = \begin{bmatrix} 1 & -1 & 2 \\ -2 & 1 & 3 \\ 3 & 2 & -3 \end{bmatrix}$

The characteristic equation is $|A - \lambda I| = 0$

$$\begin{vmatrix} 1-\lambda & -1 & 2 \\ -2 & 1-\lambda & 3 \\ 3 & 2 & -3-\lambda \end{vmatrix} = 0$$

$(1-\lambda)[(1-\lambda)(-3-\lambda) - 6] + 1[-2(-3-\lambda) - 9] + 2[-4 - 3(1-\lambda)] = 0$

$-\lambda^3 - \lambda^2 + 19\lambda - 26 = 0 \quad \rightarrow \quad \lambda^3 + \lambda^2 - 19\lambda + 26 = 0$

$\rightarrow A3 + A2 - 19A + 26I = 0 \quad \rightarrow$ (1)

$$A^2 = AA = \begin{bmatrix} 1 & -1 & 2 \\ -2 & 1 & 3 \\ 3 & 2 & -3 \end{bmatrix}\begin{bmatrix} 1 & -1 & 2 \\ -2 & 1 & 3 \\ 3 & 2 & -3 \end{bmatrix} = \begin{bmatrix} 9 & 2 & -7 \\ 5 & 9 & -10 \\ -10 & -7 & 21 \end{bmatrix}$$

$$A^3 = A^2 A = \begin{bmatrix} 9 & 2 & -7 \\ 5 & 9 & -10 \\ -10 & -7 & 21 \end{bmatrix}\begin{bmatrix} 1 & -1 & 2 \\ -2 & 1 & 3 \\ 3 & 2 & -3 \end{bmatrix}$$

$$= \begin{bmatrix} -16 & -21 & 45 \\ -43 & -16 & 67 \\ 67 & 45 & -104 \end{bmatrix}$$

$$(1) \rightarrow \begin{bmatrix} -16 & -21 & 45 \\ -43 & -16 & 67 \\ 67 & 45 & -104 \end{bmatrix} + \begin{bmatrix} 9 & 2 & -7 \\ 5 & 9 & -10 \\ -10 & -7 & 21 \end{bmatrix} -$$

$$\begin{bmatrix} 19 & -19 & 38 \\ -38 & 19 & 57 \\ 57 & 38 & -57 \end{bmatrix} + \begin{bmatrix} 26 & 0 & 0 \\ 0 & 26 & 0 \\ 0 & 0 & 26 \end{bmatrix} = 0$$

Hence Cayley – Hamilton theorem is satisfied.

(1) X by A^{-1} ➜ $A^2 + A - 19I + 26A^{-1} = 0$

$$A^{-1} = \frac{1}{26}[-A^2 - A + 19I]$$

$$= \left[\begin{bmatrix} -9 & -2 & 7 \\ -5 & -9 & 10 \\ 10 & 7 & -21 \end{bmatrix} - \begin{bmatrix} 1 & -1 & 2 \\ -2 & 1 & 3 \\ 3 & 2 & -3 \end{bmatrix} + \begin{bmatrix} 19 & 0 & 0 \\ 0 & 19 & 0 \\ 0 & 0 & 19 \end{bmatrix} \right]$$

$$= \frac{1}{26} \begin{bmatrix} 9 & -1 & 5 \\ -3 & 9 & 7 \\ 7 & 5 & 1 \end{bmatrix}.$$

3) Show that the matrix $A = \begin{bmatrix} a+ic & -b+id \\ b+id & a-ic \end{bmatrix}$ is unitary if and only if $a^2 + b^2 + c^2 + d^2 = 1$.

Sol: Given: $A = \begin{bmatrix} a+ic & -b+id \\ b+id & a-ic \end{bmatrix}$ $\qquad \overline{A} = \begin{bmatrix} a+ic & b+id \\ -b+id & a-ic \end{bmatrix}$

$$A^* = \begin{bmatrix} a-ic & b-id \\ -b-id & a+ic \end{bmatrix}$$

$$AA^* = \begin{bmatrix} a+ic & -b+id \\ b+id & a-ic \end{bmatrix}\begin{bmatrix} a-ic & b-id \\ -b-id & a+ic \end{bmatrix}$$

$$= \begin{bmatrix} a^2 + b^2 + c^2 + d^2 & 0 \\ 0 & a^2 + b^2 + c^2 + d^2 \end{bmatrix}$$

$$= (a^2 + b^2 + c^2 + d^2) \begin{bmatrix} 1 & 0 \\ 0 & 1 \end{bmatrix} = I$$

Therefore, A is unitary.

4) Find the rank of the matrix $\begin{bmatrix} 1 & 1 & 1 & -1 \\ 1 & 2 & 3 & 4 \\ 3 & 4 & 5 & 2 \end{bmatrix}$.

Sol: Let $A = \begin{bmatrix} 1 & 1 & 1 & -1 \\ 1 & 2 & 3 & 4 \\ 3 & 4 & 5 & 2 \end{bmatrix}$ $(R_3 = R_3 - (2R_1 + R_2))$

$$\begin{bmatrix} 1 & 1 & 1 & -1 \\ 1 & 2 & 3 & 4 \\ 0 & 0 & 0 & 0 \end{bmatrix} (R_2 = R_2 - (R_1 R_2))$$

$$\begin{bmatrix} 1 & 1 & 1 & -1 \\ 0 & 0 & 0 & 8 \\ 0 & 0 & 0 & 0 \end{bmatrix}$$

Therefore, Rank of A = 2.

5) Show that the matrix $A = \begin{bmatrix} \dfrac{1+i}{2} & \dfrac{-1+i}{2} \\ \dfrac{1+i}{2} & \dfrac{1-i}{2} \end{bmatrix}$ is unitary.

Sol: Given: $A = \begin{bmatrix} \dfrac{1+i}{2} & \dfrac{-1+i}{2} \\ \dfrac{1+i}{2} & \dfrac{1-i}{2} \end{bmatrix}$ $A^| = \begin{bmatrix} \dfrac{1+i}{2} & \dfrac{1+i}{2} \\ \dfrac{-1+i}{2} & \dfrac{1-i}{2} \end{bmatrix}$

$$A^* = \begin{bmatrix} \dfrac{1-i}{2} & \dfrac{1-i}{2} \\[2mm] \dfrac{-1-i}{2} & \dfrac{1+i}{2} \end{bmatrix}$$

$$A^*A = \begin{bmatrix} \dfrac{1-i}{2} & \dfrac{1-i}{2} \\[2mm] \dfrac{-1-i}{2} & \dfrac{1+i}{2} \end{bmatrix}\begin{bmatrix} \dfrac{1+i}{2} & \dfrac{-1+i}{2} \\[2mm] \dfrac{1+i}{2} & \dfrac{1-i}{2} \end{bmatrix}$$

$$= \begin{bmatrix} \dfrac{1-i^2}{4}+\dfrac{1-i^2}{4} & \dfrac{1-i}{2} \\[2mm] \dfrac{-(1+i)^2}{4}+\dfrac{(1+i)^2}{4} & \dfrac{1+i}{2} \end{bmatrix}\begin{bmatrix} \dfrac{1+i}{2} & \dfrac{-1+i}{2} \\[2mm] \dfrac{1+i}{2} & \dfrac{1-i}{2} \end{bmatrix}$$

$$= \begin{bmatrix} \dfrac{1-i^2}{4}+\dfrac{1-i^2}{4} & \dfrac{-(1-i)^2}{4}+\dfrac{(1-i)^2}{4} \\[2mm] \dfrac{-(1+i)^2}{4}+\dfrac{(1+i)^2}{4} & \dfrac{1-i^2}{4}+\dfrac{1-i^2}{4} \end{bmatrix}$$

$$= \begin{bmatrix} 1 & 0 \\ 0 & 1 \end{bmatrix} = I$$

Therefore, A is unitary.

6) Find the rank of the matrix $A = \begin{bmatrix} 1 & 2 & 3 & 2 \\ 2 & 4 & 6 & 4 \\ 3 & 4 & 5 & 2 \end{bmatrix}$.

Sol: Given: $A = \begin{bmatrix} 1 & 2 & 3 & 2 \\ 2 & 4 & 6 & 4 \\ 3 & 4 & 5 & 2 \end{bmatrix}$ $(R_3 = R_3 - (2+R_1))$

$$\begin{bmatrix} 1 & 2 & 3 & 2 \\ 2 & 4 & 6 & 4 \\ 0 & 0 & 0 & -2 \end{bmatrix}$$

$$(R_2 = R_2 - 2R_1) \begin{bmatrix} 1 & 2 & 3 & 2 \\ 0 & 0 & 0 & 0 \\ 0 & 0 & 0 & -2 \end{bmatrix}$$

Therefore, Rank of A = 2.

7) Express $\begin{bmatrix} 6 & 8 & 5 \\ 4 & 2 & 3 \\ 9 & 7 & 1 \end{bmatrix}$ as sum of a symmetric and skew – symmetric matrix.

Sol: Let $A = \begin{bmatrix} 6 & 8 & 5 \\ 4 & 2 & 3 \\ 9 & 7 & 1 \end{bmatrix}$ $A^| = \begin{bmatrix} 6 & 4 & 9 \\ 8 & 2 & 7 \\ 5 & 3 & 1 \end{bmatrix}$

$$\frac{1}{2}(A + A^|) = \begin{bmatrix} 6 & 6 & 7 \\ 6 & 2 & 5 \\ 7 & 5 & 1 \end{bmatrix}$$

$$\frac{1}{2}(A + A^|) = \begin{bmatrix} 0 & 2 & -2 \\ -2 & 0 & -2 \\ 2 & 2 & 0 \end{bmatrix}$$

We know that $A = \frac{1}{2}(A + A^|) + \frac{1}{2}(A - A^|)$

$$= \begin{bmatrix} 6 & 6 & 7 \\ 6 & 2 & 5 \\ 7 & 5 & 1 \end{bmatrix} \begin{bmatrix} 0 & 2 & -2 \\ -2 & 0 & -2 \\ 2 & 2 & 0 \end{bmatrix}.$$

8) Verify Cayley – Hamilton Theorem for the matrix $A = \begin{bmatrix} 7 & 3 \\ 2 & 6 \end{bmatrix}$ and hence find A^{-1}.

Sol: Given $A = \begin{bmatrix} 7 & 3 \\ 2 & 6 \end{bmatrix}$

The characteristic equation is $|A - \lambda I| = 0$

$$\begin{vmatrix} 7-\lambda & 3 \\ 2 & 6-\lambda \end{vmatrix} = 0$$

$$\lambda^2 - 13\lambda + 36 = 0$$

$$A^2 - 13A + 36 = 0 \quad \Rightarrow \qquad\qquad (1)$$

$$A^2 = \begin{bmatrix} 7 & 3 \\ 2 & 6 \end{bmatrix}\begin{bmatrix} 7 & 3 \\ 2 & 6 \end{bmatrix} = \begin{bmatrix} 55 & 39 \\ 26 & 42 \end{bmatrix}$$

$$(1) \Rightarrow \quad \begin{bmatrix} 55 & 39 \\ 26 & 42 \end{bmatrix} - \begin{bmatrix} 91 & 39 \\ 26 & 42 \end{bmatrix} + \begin{bmatrix} 36 & 0 \\ 0 & 36 \end{bmatrix} = 0$$

Hence Cayley – Hamilton theorem is verified.

(1) X by A^{-1} $\Rightarrow$ $A - 13I + 36A^{-1} = 0$

$$A^{-1} = \frac{1}{36}[13I - A] = \frac{1}{36}\begin{bmatrix} 6 & -2 \\ -3 & 7 \end{bmatrix}.$$

9)   Show that every square matrix can be uniquely expressed as the sum of a symmetric and skew – symmetric matrix.

Sol: Let A be square matrix.

$$A = \frac{1}{2}(A + A^{|}) + \frac{1}{2}(A - A^{|}) \quad \Rightarrow \qquad\qquad (1)$$

Now $(A + A^{|})^{|} = A^{|} + (A^{|})^{|} = A^{|} + A = A + A^{|}$

Now $(A - A^{|})^{|} = A^{|} - (A^{|})^{|} = A^{|} - A = -(A - A^{|})$

Therefore, $A + A^{|}$ is symmetric and $A - A^{|}$ is skew – symmetric.

From (1) A = P + Q

Where $P = \frac{1}{2}(A + A^{|})$ is symmetric

And $Q = \frac{1}{2}(A - A^{|})$ is skew – symmetric.

Therefore, any square matrix can be uniquely expressed as the sum of a symmetric and skew – symmetric matrix.

10) Verify Cayley – Hamilton Theorem for the matrix $A = \begin{bmatrix} 1 & 2 \\ 2 & -1 \end{bmatrix}$ and hence find A^{-1}.

Sol: Given: $A = \begin{bmatrix} 1 & 2 \\ 2 & -1 \end{bmatrix}$

The characteristic equation is $|A - \lambda I| = 0$

$$\begin{vmatrix} 1-\lambda & 2 \\ 2 & -1-\lambda \end{vmatrix} = 0$$

$$\lambda^2 - 5 = 0$$

$$A^2 - 5I = 0 \quad \rightarrow \qquad (1)$$

$$A^2 = \begin{bmatrix} 1 & 2 \\ 2 & -1 \end{bmatrix}\begin{bmatrix} 1 & 2 \\ 2 & -1 \end{bmatrix} = \begin{bmatrix} 5 & 0 \\ 0 & 5 \end{bmatrix}$$

$$(1) \rightarrow \begin{bmatrix} 5 & 0 \\ 0 & 5 \end{bmatrix} - \begin{bmatrix} 5 & 0 \\ 0 & 5 \end{bmatrix} = 0$$

Hence Cayley – Hamilton theorem is verified.

(1) X by A^{-1} $\rightarrow$ $A - 5A^{-1} = 0$

$$A^{-1} = \frac{1}{5}[A] = \frac{1}{5}\begin{bmatrix} 1 & 2 \\ 2 & -1 \end{bmatrix}.$$

11) Test for consistency and solve if consistent.

$x + 2y - z = 3$; $3x - y + 2z = 1$; $2x - 2y + 3z = 2$; $x - y + z = -1$.

Sol: Given: $x + 2y - z = 3$; $3x - y + 2z = 1$; $2x - 2y + 3z = 2$; $x - y + z = -1$.

Let $A = \begin{bmatrix} 1 & 2 & -1 \\ 3 & -1 & 2 \\ 2 & -2 & 3 \\ 1 & -1 & 1 \end{bmatrix}$ $B = \begin{bmatrix} 3 \\ 1 \\ 2 \\ -1 \end{bmatrix}$

Then the augmented matrix is $(A,B) = \begin{bmatrix} 1 & 2 & -1 & 3 \\ 3 & -1 & 2 & 1 \\ 2 & -2 & 3 & 2 \\ 1 & -1 & 1 & -1 \end{bmatrix}$

$$= \begin{bmatrix} 1 & 2 & -1 & 3 \\ 0 & -7 & 5 & -8 \\ 0 & -6 & 5 & -4 \\ 0 & -3 & 2 & -4 \end{bmatrix} \left(R_1 R_1, R_2 = R_2 - 3R_1, R_3 = R_3 - 2R_1, R_4, = R_4 - R_1\right)$$

$$= \begin{bmatrix} 1 & 2 & -1 & 3 \\ 0 & -1 & 0 & -4 \\ 0 & -6 & 5 & -4 \\ 0 & -3 & 2 & -4 \end{bmatrix} \left(R_1, R_2 = R_2 - R_3, R_3 R_4\right)$$

$$= \begin{bmatrix} 1 & 2 & -1 & 3 \\ 0 & -1 & 0 & -4 \\ 0 & 0 & 5 & 20 \\ 0 & 0 & 2 & 8 \end{bmatrix} \left(R_1, R_2, R_3 = R_3 - 6R_2, R_4 = R_4 - 3R_2\right)$$

$$= \begin{bmatrix} 1 & 2 & -1 & 3 \\ 0 & -1 & 0 & -4 \\ 0 & 0 & 1 & 4 \\ 0 & 0 & 1 & 4 \end{bmatrix} \left(R_1, R_2, R_3 = R3/5, R_4 = R_4/2\right)$$

$$= \begin{bmatrix} 1 & 2 & -1 & 3 \\ 0 & -1 & 0 & -4 \\ 0 & 0 & 1 & 4 \\ 0 & 0 & 0 & 0 \end{bmatrix} \left(R_1, R_2, R_3, R_4 = R_4 - R_3\right)$$

Therefore, $\rho\,(A\,,B) = 3$

Therefore, the given system of equations is consistent and has a unique solution.

$Z = 4$, $y = 4$, $x = -1$.

10 MARKS

1) Show that the equation $x + y + z = 6$; $x + 2y + 3z = 14$; $x + 4y + 7z = 30$ are consistent and solve them.

Sol: Let $A = \begin{bmatrix} 1 & 1 & 1 \\ 1 & 2 & 3 \\ 1 & 4 & 7 \end{bmatrix}$ $\qquad B = \begin{bmatrix} 6 \\ 14 \\ 30 \end{bmatrix}$

$$(A, B) = \begin{bmatrix} 1 & 1 & 1 & 6 \\ 1 & 2 & 3 & 14 \\ 1 & 4 & 7 & 30 \end{bmatrix}$$

$$\begin{bmatrix} 1 & 1 & 1 & 6 \\ 0 & 1 & 2 & 8 \\ 0 & 3 & 6 & 24 \end{bmatrix} \quad (R_1, R_2 = R_2 - R_1, R_3 = R_3 - R_1)$$

$$\begin{bmatrix} 1 & 1 & 1 & 6 \\ 0 & 1 & 2 & 8 \\ 0 & 0 & 0 & 0 \end{bmatrix} \quad (R_1, R_2, R_3 = R_3 - 3R_2)$$

Therefore, $\rho\,(A\,,B) = \rho\,(A) = 2 < 3$

Hence the given equations are consistent. But there are infinite number of solutions.

Also $x + y + z = 6$

y + 2z = 8

y = 8 – 2z

x = 6 – (8 – 2z) – z = z – 2

Taking z = k, where k is arbitrary, the solutions are x = k – 2 , y = 8 – 2k, z = k.

2) Show that the matrix $\begin{bmatrix} 0 & c & -b \\ -c & 0 & a \\ b & -a & 0 \end{bmatrix}$ satisfies Cayley – Hamilton Theorem.

Sol: Let $A = \begin{bmatrix} 0 & c & -b \\ -c & 0 & a \\ b & -a & 0 \end{bmatrix}$

The characteristic equation is $|A - \lambda I| = 0$

$$\begin{bmatrix} -\lambda & c & -b \\ -c & -\lambda & a \\ b & -a & -\lambda \end{bmatrix} = 0$$

$$-\lambda(\lambda^2 + a^2) - c(c\lambda - ab) - b(ac + b\lambda) = 0$$

$$-\lambda^3 - \lambda a^2 - c^2\lambda + abc - abc - b^2\lambda = 0$$

$$\lambda^3 + \lambda(a^2 + b^2 + c^2) = 0$$

$$A^3 + (a^2 + b^2 + c^2)A = 0$$

$$A^2 = AA = \begin{bmatrix} 0 & c & -b \\ -c & 0 & a \\ b & -a & 0 \end{bmatrix}\begin{bmatrix} 0 & c & -b \\ -c & 0 & a \\ b & -a & 0 \end{bmatrix} = $$

$$\begin{bmatrix} -c^2 - b^2 & ab & ac \\ ac & -c^2 - a^2 & bc \\ ac & bc & -b^2 - a^2 \end{bmatrix}$$

$$A^3 = A^2 A = \begin{bmatrix} -c^2 - b^2 & ab & ac \\ ac & -c^2 - a^2 & bc \\ ac & bc & -b^2 - a^2 \end{bmatrix} \begin{bmatrix} 0 & c & -b \\ -c & 0 & a \\ b & -a & 0 \end{bmatrix}$$

$$= -(a^2 + b^2 + c^2)A$$

Therefore, $A^3 + (a^2 + b^2 + c^2)A = 0$

3) Verify Cayley – Hamilton Theorem for the matrix $A = \begin{bmatrix} 2 & -1 & 1 \\ -1 & 2 & -1 \\ 1 & -1 & 2 \end{bmatrix}$ and hence find A^{-1}

Sol: Given: $A = \begin{bmatrix} 2 & -1 & 1 \\ -1 & 2 & -1 \\ 1 & -1 & 2 \end{bmatrix}$

The characteristic equation is $|A - \lambda I| = 0$

$$\begin{bmatrix} 2-\lambda & -1 & 1 \\ -1 & 2-\lambda & -1 \\ 1 & -1 & 2-\lambda \end{bmatrix} = 0$$

$(2 - \lambda)[(2 - \lambda)^2 - 1] + 1[-2 + \lambda + 1] + [1 - 2 + \lambda] = 0$

$(2 - \lambda)(3 - 4\lambda + \lambda^2) + (\lambda - 1) + (\lambda - 1) = 0$

$6 - 8\lambda + 2\lambda^2 - 3\lambda + 4\lambda^2 - \lambda^3 + 2\lambda - 2 = 0$

$-\lambda^3 + 6\lambda^2 - 9\lambda + 4 = 0$

$\lambda^3 - 6\lambda^2 + 9\lambda - 4 = 0$

Now $A^3 - 6A^2 + 9A - 4I = 0$

$$A^2 = \begin{bmatrix} 2 & -1 & 1 \\ -1 & 2 & -1 \\ 1 & -1 & 2 \end{bmatrix} \begin{bmatrix} 2 & -1 & 1 \\ -1 & 2 & -1 \\ 1 & -1 & 2 \end{bmatrix} = \begin{bmatrix} 6 & -5 & 5 \\ -5 & 6 & -5 \\ 5 & -5 & 6 \end{bmatrix}$$

$$A^3 = A^2 A = \begin{bmatrix} 6 & -5 & 5 \\ -5 & 6 & -5 \\ 5 & -5 & 6 \end{bmatrix} \begin{bmatrix} 2 & -1 & 1 \\ -1 & 2 & -1 \\ 1 & -1 & 2 \end{bmatrix} = \begin{bmatrix} 22 & -21 & 21 \\ -21 & 22 & -21 \\ 21 & -21 & 22 \end{bmatrix}$$

$$A^3 - 6A^2 + 9A - 4I = \begin{bmatrix} 22 & -21 & 21 \\ -21 & 22 & -21 \\ 21 & -21 & 22 \end{bmatrix} - 6\begin{bmatrix} 6 & -5 & 5 \\ -5 & 6 & -5 \\ 5 & -5 & 6 \end{bmatrix} +$$

$$9\begin{bmatrix} 2 & -1 & 1 \\ -1 & 2 & -1 \\ 1 & -1 & 2 \end{bmatrix} - 4\begin{bmatrix} 1 & 0 & 0 \\ 0 & 1 & 0 \\ 0 & 0 & 1 \end{bmatrix} = \begin{bmatrix} 0 & 0 & 0 \\ 0 & 0 & 0 \\ 0 & 0 & 0 \end{bmatrix}$$

Therefore, $A^3 - 6A^2 + 9A - 4I = 0$

Multiplying by A^{-1}, we get $A^2 - 6A + 9I - 4A^{-1} = 0$

$$A^{-1} = \frac{1}{4}[A^2 - 6A + 9I]$$

$$A^{-1} = \frac{1}{4}\left[\begin{bmatrix} 6 & -5 & 5 \\ -5 & 6 & -5 \\ 5 & -5 & 6 \end{bmatrix} - 6\begin{bmatrix} 2 & -1 & 1 \\ -1 & 2 & -1 \\ 1 & -1 & 2 \end{bmatrix} + 9\begin{bmatrix} 1 & 0 & 0 \\ 0 & 1 & 0 \\ 0 & 0 & 1 \end{bmatrix}\right]$$

$$A^{-1} = \frac{1}{4}\begin{bmatrix} 3 & 1 & -1 \\ 1 & 3 & 1 \\ -1 & 1 & 3 \end{bmatrix}$$

4) Find the eigen values and eigen vectors of the matrix $\begin{bmatrix} 1 & 0 & 0 \\ 0 & 3 & -1 \\ 0 & -1 & 3 \end{bmatrix}$.

Sol: Let $A = \begin{bmatrix} 1 & 0 & 0 \\ 0 & 3 & -1 \\ 0 & -1 & 3 \end{bmatrix}$

Eigen Values are given by the equation $|A - \lambda I| = 0$

$$\begin{bmatrix} 1-\lambda & 0 & 0 \\ 0 & 3-\lambda & -1 \\ 0 & -1 & 3-\lambda \end{bmatrix} = 0$$

$(1 - \lambda)[(3 - \lambda)^2 - 1] = 0$

$(1 - \lambda)(\lambda - 2)(\lambda - 4) = 0$

$\lambda = 1, 2, 4.$

Therefore, eigen values are 1, 2, 4.

Eigen Vectors are given by the equation $(A - \lambda I)(X) = 0$

$$\begin{bmatrix} 1-\lambda & 0 & 0 \\ 0 & 3-\lambda & -1 \\ 0 & -1 & 3-\lambda \end{bmatrix}\begin{bmatrix} x_1 \\ x_2 \\ x_3 \end{bmatrix} = 0$$

$(1 - \lambda)x_1 = 0$

$(3 - \lambda)x_2 - x_3 = 0$

$(3 - \lambda)x_3 - x_2 = 0$

CASE 1 : $\lambda = 1$

$x_1 = 0$

$2x_2 - x_3 = 0$ ➜   (1)

$2x_3 - x_2 = 0$ ➜   (2)

(2)*2 ➜ $4x_3 - 2x_2 = 0$ ➜   (3)

(2) + (3) ➜ $3x_3 = 0$ ➜ $x_3 = 0$

From (1), $x_2 = 0$

Eigen Vector $= \begin{bmatrix} 0 \\ 0 \\ 0 \end{bmatrix}$

CASE 2 : $\lambda = 2$

$$-x_1 = 0 \rightarrow x_1 = 0$$

$$x_2 - x_3 = 0 \rightarrow \tag{1}$$

$$-x_2 + x_3 = 0 \rightarrow \tag{2}$$

(1) and (2) are same. i.e., $x_2 = x_3 = k$ (say)

$$\text{Eigen Vector} = \begin{bmatrix} 0 \\ k \\ k \end{bmatrix} \text{ or } \begin{bmatrix} 0 \\ 1 \\ 1 \end{bmatrix}$$

CASE 3 : $\lambda = 4$

$$-3x_1 = 0 \rightarrow x_1 = 0$$

$$-x_2 - x_3 = 0 \rightarrow \tag{1}$$

$$-x_2 - x_3 = 0 \rightarrow \tag{2}$$

(1) and (2) are same. i.e., $x_2 = -x_3 = k$ (say)

$$\text{Eigen Vector} = \begin{bmatrix} 0 \\ k \\ -k \end{bmatrix} \text{ or } \begin{bmatrix} 0 \\ 1 \\ -1 \end{bmatrix}$$

5) Find the eigen values and eigen vectors of the matrix $A = \begin{bmatrix} 2 & 0 & 1 \\ 0 & 2 & 0 \\ 1 & 0 & 2 \end{bmatrix}$.

Sol: Given: $A = \begin{bmatrix} 2 & 0 & 1 \\ 0 & 2 & 0 \\ 1 & 0 & 2 \end{bmatrix}$.

Eigen Values are given by the equation $|A - \lambda I| = 0$

$$\begin{bmatrix} 2-\lambda & 0 & 1 \\ 0 & 2-\lambda & 0 \\ 1 & 0 & 2-\lambda \end{bmatrix} = 0$$

$$(2 - \lambda)[(2 - \lambda)^2] = 0$$

$$(2 - \lambda)^3 = 0$$

$$8 - 12\lambda + 6\lambda^2 - \lambda^3 = 0$$

$$\lambda^3 - 6\lambda^2 + 12\lambda - 8 = 0$$

$$\lambda = 2, 2, 2.$$

Therefore, eigen values are 2, 2, 2.

Eigen Vectors are given by the equation $(A - \lambda I)(X) = 0$

$$\begin{bmatrix} 2-\lambda & 0 & 1 \\ 0 & 2-\lambda & 0 \\ 1 & 0 & 2-\lambda \end{bmatrix} \begin{bmatrix} x_1 \\ x_2 \\ x_3 \end{bmatrix} = 0$$

$$(2 - \lambda)x_1 + x_3 = 0$$

$$(2 - \lambda)x_2 = 0$$

$$x_1 + (2 - \lambda)x_3 = 0$$

$$\lambda = 2$$

$$x_1 = 0, \ x_2 = 0, \ x_3 = 0$$

Eigen Vector $= \begin{bmatrix} 0 \\ 0 \\ 0 \end{bmatrix}$.

6) Find the characteristic equation of the matrix $A = \begin{bmatrix} 1 & 1 & 3 \\ 5 & 2 & 6 \\ -2 & -1 & -3 \end{bmatrix}$ and show that the matrix A satisfies the equation.

Sol: Given: $A = \begin{bmatrix} 1 & 1 & 3 \\ 5 & 2 & 6 \\ -2 & -1 & -3 \end{bmatrix}$

The characteristic equation is $|A - \lambda I| = 0$

$$\begin{bmatrix} 1-\lambda & 1 & 3 \\ 5 & 2-\lambda & 6 \\ -2 & -1 & -3-\lambda \end{bmatrix} = 0$$

$$(1 - \lambda)[-(2 - \lambda)(3 + \lambda) + 6] - 1[-5(3 + \lambda) + 12] + 3[-5 + 4 - 2\lambda] = 0$$

$$(1 - \lambda)(\lambda^2 + \lambda) + 5\lambda + 3 - 6\lambda - 3 = 0$$

$\lambda^2 + \lambda^3 - \lambda^2 + 5\lambda + 3 - 6\lambda - 3 = 0 \quad \Rightarrow \quad \lambda^3 = 0$

Now we will show that the matrix satisfies the characteristic equation. For that we have to verify that $A^3 = 0$

$$A^2 = AA = \begin{bmatrix} 1 & 1 & 3 \\ 5 & 2 & 6 \\ -2 & -1 & -3 \end{bmatrix}\begin{bmatrix} 1 & 1 & 3 \\ 5 & 2 & 6 \\ -2 & -1 & -3 \end{bmatrix} = \begin{bmatrix} 0 & 0 & 0 \\ 3 & 3 & 9 \\ -1 & -1 & -3 \end{bmatrix}$$

$$A^3 = A^2A = \begin{bmatrix} 0 & 0 & 0 \\ 3 & 3 & 9 \\ -1 & -1 & -3 \end{bmatrix}\begin{bmatrix} 1 & 1 & 3 \\ 5 & 2 & 6 \\ -2 & -1 & -3 \end{bmatrix} = \begin{bmatrix} 0 & 0 & 0 \\ 0 & 0 & 0 \\ 0 & 0 & 0 \end{bmatrix}$$

Therefore, $A^3 = 0$.

7) Using Cayley – Hamilton Theorem, find A^4, for the matrix $A = \begin{bmatrix} 2 & -2 & 1 \\ 0 & 1 & 2 \\ 1 & 0 & 1 \end{bmatrix}$.

Sol: Given: $A = \begin{bmatrix} 2 & -2 & 1 \\ 0 & 1 & 2 \\ 1 & 0 & 1 \end{bmatrix}$

The characteristic equation is $|A - \lambda I| = 0$

$$\begin{vmatrix} 2-\lambda & -2 & 1 \\ 0 & 1-\lambda & 2 \\ 1 & 0 & 1-\lambda \end{vmatrix} = 0$$

$(2 - \lambda)[(1 - \lambda)(-1 - \lambda) - 0] + 2[-2] + 1[-(1 - \lambda)] = 0$

$-\lambda^3 + 4\lambda^2 - 4\lambda - 3 = 0 \quad \Rightarrow \quad \lambda^3 - 4\lambda^2 + 4\lambda + 3 = 0$

$\Rightarrow A^3 - 4A^2 + 4A + 3I = 0 \quad \Rightarrow \quad \hspace{4cm} (1)$

$$A^2 = AA = \begin{bmatrix} 2 & -2 & 1 \\ 0 & 1 & 2 \\ 1 & 0 & 1 \end{bmatrix}\begin{bmatrix} 2 & -2 & 1 \\ 0 & 1 & 2 \\ 1 & 0 & 1 \end{bmatrix} = \begin{bmatrix} 5 & -6 & -1 \\ 2 & 1 & 4 \\ 3 & -2 & 2 \end{bmatrix}$$

$$A^3 = A^2 A = \begin{bmatrix} 5 & -6 & -1 \\ 2 & 1 & 4 \\ 3 & -2 & 2 \end{bmatrix} \begin{bmatrix} 2 & -2 & 1 \\ 0 & 1 & 2 \\ 1 & 0 & 1 \end{bmatrix} = \begin{bmatrix} 9 & -16 & -8 \\ 8 & -3 & 8 \\ 8 & -8 & 1 \end{bmatrix}$$

$$(1) \Rightarrow \begin{bmatrix} 9 & -16 & -8 \\ 8 & -3 & 8 \\ 8 & -8 & 1 \end{bmatrix} - \begin{bmatrix} 20 & -24 & -4 \\ 8 & 4 & 16 \\ 12 & -8 & 8 \end{bmatrix} + \begin{bmatrix} 8 & -8 & 4 \\ 0 & 4 & 8 \\ 4 & 0 & 4 \end{bmatrix} + \begin{bmatrix} 3 & 0 & 0 \\ 0 & 3 & 0 \\ 0 & 0 & 3 \end{bmatrix} = 0$$

Hence Cayley – Hamilton theorem is satisfied.

(1) X by A^4 $\Rightarrow$ $A^4 - 4A^3 + 4A^2 + 3A = 0$

$A^4 = 4A^3 - 4A^2 - 3A$

$$= \left[\begin{bmatrix} 36 & -64 & -32 \\ 32 & -12 & 32 \\ 32 & -32 & 4 \end{bmatrix} - \begin{bmatrix} 20 & -24 & -4 \\ 8 & 4 & 16 \\ 12 & -8 & 8 \end{bmatrix} - \begin{bmatrix} 6 & -6 & 3 \\ 0 & 3 & 6 \\ 3 & 0 & 3 \end{bmatrix} \right]$$

$$= \begin{bmatrix} 10 & -34 & -25 \\ 24 & -19 & 10 \\ 17 & -24 & -7 \end{bmatrix}.$$

UNIT – 4

DIFFERENTIAL CALCULUS

2 MARKS

1) Find the radius of curvature at x = 0 on $y = e^x$.

Sol: Given: $y = e^x$ ➜   (1)

Since x = 0, we have $y = e^0$ ➜ y = 1

Therefore, we have to find the radius of curvature at (0,1) on $y = e^x$.

Differentiating (1) with respect to x, we get

$y_1 = e^x$ ➜ (2) [where $y_1 = \dfrac{dy}{dx}$]

Again differentiating (2) with respect to x, we get

$y_2 = e^x$ [where $y_1 = y_1 = \dfrac{d^2 y}{dx^2}$]

Now at the point (0,1) we have $y_1 = 1$, $y_2 = 1$.

We know that $\rho = \dfrac{\left[1+\left[\dfrac{dy}{dx}\right]^2\right]^{\frac{3}{2}}}{\dfrac{d^2 y}{dx^2}} = \left[1+1\right]^{\frac{3}{2}} = 2\sqrt{2}.$

2) If $x = r \cos\theta$, $y = r \sin\theta$, then find $\dfrac{\partial(x, y)}{\partial(r, \theta)}$.

Sol: Given: $x = r \cos\theta$, $y = r \sin\theta$

$$\frac{\partial x}{\partial r} = cos\theta, \frac{\partial y}{\partial r} = sin\theta \;\&\; \frac{\partial x}{\partial \theta} = -r\,sin\theta, \frac{\partial y}{\partial \theta} = r\,cos\theta$$

$$\text{Now } \frac{\partial(x, y)}{\partial(r, \theta)} = \begin{vmatrix} \dfrac{\partial x}{\partial r} & \dfrac{\partial x}{\partial \theta} \\ \dfrac{\partial y}{\partial r} & \dfrac{\partial y}{\partial \theta} \end{vmatrix} = \begin{vmatrix} \cos\theta & -r\sin\theta \\ \sin\theta & r\cos\theta \end{vmatrix} = r\cos^2\theta + r\sin^2\theta = r$$

3) Give the formula for the radius of curvature in Cartesian form.

Sol: $\rho = \dfrac{\left[1 + \left[\dfrac{dy}{dx}\right]^2\right]^{\frac{3}{2}}}{\dfrac{d^2 y}{dx^2}}$ where $\dfrac{d^2 y}{dx^2} \neq 0$.

4) If $x = u(1+v)$ and $y = v(1+u)$, then find $\dfrac{\partial(x, y)}{\partial(u, v)}$.

Sol: Given: $x = u(1+v)$ and $y = v(1+u)$

$$\frac{\partial x}{\partial u} = 1 + v, \frac{\partial y}{\partial u} = v \And \frac{\partial x}{\partial v} = u, \frac{\partial y}{\partial v} = 1 + u$$

$$\text{Now } \frac{\partial(x, y)}{\partial(u, v)} = \begin{vmatrix} \dfrac{\partial x}{\partial u} & \dfrac{\partial x}{\partial v} \\ \dfrac{\partial y}{\partial u} & \dfrac{\partial y}{\partial v} \end{vmatrix} = \begin{vmatrix} 1+v & u \\ v & 1+u \end{vmatrix} = (1+u)(1+v) - uv = 1 + u + v.$$

5) Find $\dfrac{\partial(u, v)}{\partial(x, y)}$ if $u = x^2$, $v = y^2$.

Sol: Given: $u = x^2$, $v = y^2$

$$\frac{\partial u}{\partial x} = 2x, \frac{\partial u}{\partial y} = 0 \And \frac{\partial v}{\partial x} = 0, \frac{\partial v}{\partial y} = 2y$$

$$\text{Now, } \frac{\partial(u, v)}{\partial(x, y)} = \begin{vmatrix} \dfrac{\partial u}{\partial x} & \dfrac{\partial u}{\partial y} \\ \dfrac{\partial v}{\partial x} & \dfrac{\partial v}{\partial y} \end{vmatrix} = \begin{vmatrix} 2x & 0 \\ 0 & 2y \end{vmatrix} = 4xy.$$

6)   Find the nth derivative of e^{ax}.

Sol: $y = e^{ax}$

$$y_1 = e^{ax}.a$$

$$y_2 = e^{ax}.a^2$$

$$y_3 = e^{ax}.a^3$$

Therefore, $y_n = a^n e^{ax}$

7)   Define Curvature.

Sol: The direction of the curve at a point is determined by the slope at that point, but a measure of the rapidity with which the curve is bending at the point is given by the rate of change of with respect to s, the arc length. This rate of change is called the curvature of the curve at the particular point. i.e., $\lim\limits_{\delta s \to 0}\left[\dfrac{\delta \varphi}{\delta s}\right] = \dfrac{d\varphi}{ds}$.

8)   Find the radius of curvature of the ellipse $x = a \cos\,\theta$, $y = b \sin\,\theta$, at $\theta = \dfrac{\pi}{4}$.

Sol: Given: $x = a \cos\,\theta$, $y = b \sin\,\theta$

$$x = a \cos\,\theta$$

$$\rightarrow \frac{dx}{d\theta} = -a\,sin\theta$$

$$y = b \sin\,\theta$$

$$\rightarrow \frac{dy}{d\theta} = b\,cos\theta$$

$$y_1 = \frac{dy}{dx} = \frac{\dfrac{dy}{d\theta}}{\dfrac{dx}{d\theta}} = \frac{b\,cos\theta}{-a\,sin\theta} = -\frac{b}{a}\,cot\theta$$

$$y_2 = \frac{d^2y}{dx^2} = -\frac{b}{a^2}\,cosec^2\theta$$

We know that $\rho = \dfrac{\left[1+\left[\dfrac{dy}{dx}\right]^2\right]^{\frac{3}{2}}}{\dfrac{d^2y}{dx^2}}$

$$= \frac{\left[1+\dfrac{b^2}{a^2}cot^2\theta\right]^{\frac{3}{2}}}{-\dfrac{b}{a^2}cosec^3\theta}$$

$$= [1+\frac{b^2}{a^2}\frac{cos^2\vartheta}{sin^2\theta}]^{\frac{3}{2}}\frac{-a^2}{b\,cosec^3\theta}$$

$$= \frac{(a^2sin^2\theta + b^2cos^2\theta)^{\frac{3}{2}}}{a^2sin^3\theta}\frac{-a^2}{b\,cosec^3\theta}$$

$$= \frac{(a^2sin^2\theta + b^2cos^2\theta)^{\frac{3}{2}}}{ab}$$

At $\theta = \dfrac{\pi}{4}, \rho = \dfrac{(\dfrac{a^2+b^2}{2})^{\frac{3}{2}}}{ab}$.

5 MARKS

---

1) If $y = \sin(m \sin^{-1} x)$, then prove that $(1 - x^2)y_{n+2} - (2n + 1)xy_{n+1} - (n^2 - m^2)y_n = 0$.

Sol: Given: $y = \sin(m \sin^{-1} x)$

Then $\sin^{-1} y = m \sin^{-1} x$

Differentiating with respect to x, we get

$$\frac{1}{\sqrt{1-y^2}} \cdot y_1 = m \cdot \frac{1}{\sqrt{1-x^2}}$$

$$\sqrt{1-x^2} \cdot y_1 = m\sqrt{1-y^2}$$

$$(1-x^2)y_1^2 = m^2(1-y^2)$$

Again differentiating with respect to x,

$$(1-x^2)2y_1y_2 + y_1^2(-2x) = m^2(-2yy_1)$$

Cancel $2y_1$ on both sides.

$$(1-x^2)y^2 - xy^1 + m^2y = 0$$

Now differentiating (2) n times with respect to x and applying Leibnitz rule, we get

$$(1-x^2)y_{n+2} - (2n+1)xy_{n+1} - (n^2 - m^2)y_n = 0.$$

2) If J_1 is the Jacobian of u(x,y) and v(x,y) and J_2 is the Jacobian of x(u,v) and y(u,v), then prove that $J_1J_2 = 1$.

Sol: $J = \dfrac{\partial(u,v)}{\partial(x,y)} = \begin{vmatrix} \dfrac{\partial u}{\partial x} & \dfrac{\partial u}{\partial y} \\[2mm] \dfrac{\partial v}{\partial x} & \dfrac{\partial v}{\partial y} \end{vmatrix}$

$J_1 = \dfrac{\partial(x,y)}{\partial(u,v)} = \begin{vmatrix} \dfrac{\partial x}{\partial u} & \dfrac{\partial x}{\partial v} \\[2mm] \dfrac{\partial y}{\partial u} & \dfrac{\partial y}{\partial v} \end{vmatrix}$

$$J_1J_2 = \begin{vmatrix} \dfrac{\partial u}{\partial x} & \dfrac{\partial u}{\partial y} \\[2mm] \dfrac{\partial v}{\partial x} & \dfrac{\partial v}{\partial y} \end{vmatrix} \begin{vmatrix} \dfrac{\partial x}{\partial u} & \dfrac{\partial x}{\partial v} \\[2mm] \dfrac{\partial y}{\partial u} & \dfrac{\partial y}{\partial v} \end{vmatrix} = \begin{vmatrix} \dfrac{\partial u}{\partial x}\dfrac{\partial x}{\partial u} + \dfrac{\partial u}{\partial y}\dfrac{\partial y}{\partial u} & \dfrac{\partial u}{\partial x}\dfrac{\partial x}{\partial v} + \dfrac{\partial u}{\partial y}\dfrac{\partial y}{\partial v} \\[2mm] \dfrac{\partial v}{\partial x}\dfrac{\partial x}{\partial u} + \dfrac{\partial v}{\partial y}\dfrac{\partial y}{\partial u} & \dfrac{\partial v}{\partial x}\dfrac{\partial x}{\partial v} + \dfrac{\partial v}{\partial y}\dfrac{\partial y}{\partial v} \end{vmatrix}$$

Let $u = f_1(x,y) \dots$ ➜ (1); $\quad x = F_1(u,v)$

Let $v = f_2(x,y) \dots$ ➜ (2); $\quad y = F_2(u,v)$

Similarly differentiating (1) partially with respect to u and v, we have

$$1 = \frac{\partial u}{\partial x}\frac{\partial x}{\partial u} + \frac{\partial u}{\partial y}\frac{\partial y}{\partial u}$$

$$0 = \frac{\partial u}{\partial x}\frac{\partial x}{\partial v} + \frac{\partial u}{\partial y}\frac{\partial y}{\partial v}$$

Similarly differentiating (2) partially with respect to u and v, we have

$$0 = \frac{\partial u}{\partial x}\frac{\partial x}{\partial u} + \frac{\partial u}{\partial y}\frac{\partial y}{\partial u}$$

$$1 = \frac{\partial v}{\partial x}\frac{\partial x}{\partial v} + \frac{\partial v}{\partial y}\frac{\partial y}{\partial v}$$

$$J_1 J_2 = \begin{vmatrix} 1 & 0 \\ 0 & 1 \end{vmatrix} = 1.$$

3)   Find the nth derivatives of $\dfrac{x^3}{(x-a)(x-b)(x-c)}$.

Sol: Let $y = \dfrac{x^3}{(x-a)(x-b)(x-c)} = 1 + \dfrac{A}{x-a} + \dfrac{B}{x-b} + \dfrac{C}{x-c}$

$x^3 = (x-a)(x-b)(x-c) + A(x-c)(x-b) +$

$B(x-a)(x-c) + c(x-a)(x-b)$

put $x = a$; $a^3 = A(a-b)(a-c)$ ➜ $A = \dfrac{a^3}{(a-b)(a-c)}$

Similarly, $B = \dfrac{b^3}{(b-a)(b-c)}$ and $C = \dfrac{c^3}{(c-a)(c-b)}$

Therefore, $y = 1 + \sum \dfrac{a^3}{(a-b)(a-c)} \dfrac{1}{x-a}$

$$y_n = \sum \frac{a^3}{(a-b)(a-c)} \frac{(-1)^n n!}{(x-a)^{n+1}}$$

4) If u = x + y + z, uv = y + z, uvw = z, then show that $\dfrac{\partial(x,y,z)}{\partial(u,v,w)} = u^2 v$.

Sol: u = x + y + z ➡ x = u − (y + z) = u − uv = u(1 − v)

y = uv − z = uv − uvw

z = uvw

$$J = \begin{vmatrix} \dfrac{\partial x}{\partial u} & \dfrac{\partial x}{\partial v} & \dfrac{\partial x}{\partial w} \\[2mm] \dfrac{\partial y}{\partial u} & \dfrac{\partial y}{\partial v} & \dfrac{\partial y}{\partial w} \\[2mm] \dfrac{\partial z}{\partial u} & \dfrac{\partial z}{\partial v} & \dfrac{\partial z}{\partial w} \end{vmatrix}$$

$$= \begin{vmatrix} 1-v & -u & 0 \\ v-vw & u-vw & -uv \\ vw & uw & uv \end{vmatrix}$$

$$= \begin{vmatrix} 1-v & -u & 0 \\ v & u & 0 \\ vw & uw & uv \end{vmatrix} (R_2 + R_3)$$

$$= u^2 v.$$

5) If $y = \dfrac{\log x}{x^2}$, then show that $x^3 y_3 + 8x^2 y_2 + 14xy_1 + 4y = 0$.

Sol: Given: $y = \dfrac{\log x}{x^2}$

$$x^2 y = \log x$$

Differentiating with respect to x,

$$x^2 y_1 + 2xy = 1/x \quad ➡ \quad x^3 y_1 + 2x^2 y = 1$$

Again differentiating with respect to x,

$$x^3y_2 + (3x^2)y_1 + (4x)y + (2x^2)y_1 = 0 \;\Rightarrow\; x^3y_2 + 5x^2y_1 + 4xy = 0$$

Once again differentiating with respect to x,

$$x^3y_3 + (3x^2)y_2 + (5x^2)y_2 + (10x)y_1 + (4x)y_1 + 4y = 0 \;\Rightarrow\;$$
$$x^3y_3 + 8x^2y_2 + 14xy_1 + 4y = 0$$

10 MARKS

1) If $y = a\cos(\log x) + b\sin(\log x)$, then show that $x^2y_{n+2} + (2n + 1)xy_{n+1} + (n^2 + 1)y_n = 0$.

Sol: Given: If $y = a\cos(\log x) + b\sin(\log x)$ $\Rightarrow$ (1)

Differentiating with respect to x, we get

$$y_1 = -a\sin(\log x).\frac{1}{x} + b\cos(\log x).\frac{1}{x}$$

$$xy_1 = -a\sin(\log x) + b\cos(\log x)$$

Again differentiating with respect to x, we get

$$xy^2 + y_1 = -a\cos(\log x).\frac{1}{x} - b\sin(\log x).\frac{1}{x}$$

$$x^2y_2 + xy_1 = [a\cos(\log x) + b\sin(\log x)]$$

$$x^2y_2 + xy_1 = -y \quad \text{from (1)}$$

$$x^2y_2 + xy_1 + y = 0$$

Now differentiating (1) n times with respect to x and applying Leibnitz rule, we get

$$x^2 y_{n+2} + (2n + 1)xy_{n+1} + (n^2 + 1)y_n = 0$$

2) If $y = e^{a\,\sin^{-1}x}$, then prove that $(1 - x^2)y_{n+2} - (2n+1)xy_{n+1} - (n^2+a^2)y_n = 0$.

Sol: Given: $y = e^{a\,\sin^{-1}x}$ ➜ (1)

Differentiating with respect to x, we get

$$y_1 = e^{a\,\sin^{-1}x} a.\frac{1}{\sqrt{1-x^2}}$$

$$\sqrt{1-x^2}\,y_1 = a\,e^{a\,\sin^{-1}x}$$

$$(1-x^2)y_1^2 = a^2\,(e^{a\,\sin^{-1}x})^2$$

$$(1-x^2)y_1^2 = a^2 y^2 \quad \text{by} \tag{1}$$

Again differentiating with respect to x, we get

$$(1-x^2)2y_1 y_2 + y_1^2(-2x) = m^2 2yy_1$$

$$(1-x^2)y_2 + xy_1 - m^2 y = 0$$

Now differentiating (1) with respect to x and applying Leibnitz rule, we get

$$(1 - x^2)y_{n+2} - (2n + 1)xy_{n+1} - (n^2 + a^2)y_n = 0$$

3) Show that the radius of curvature at (a,0) on the curve $y^2 = \dfrac{a^2(a-x)}{x}$ is $\dfrac{a}{2}$.

Sol: Given: $y^2 = \dfrac{a^2(a-x)}{x}$

Differentiating with respect to x,

$$2y\frac{dy}{dx} = \frac{x(-a)^2 - a^2(a-x)1}{x^2}$$

$$= \frac{-a^2x - a^3 + a^2x}{x^2}$$

$$2y\frac{dy}{dx} = \frac{-a^3}{x^2}$$

$$\frac{dy}{dx} = -\frac{a^3}{2yx^2}$$

$$\left[\frac{dy}{dx}\right]_{(a,0)} = \infty$$

Consider $\dfrac{dx}{dy} = -\dfrac{2yx^2}{a^3}; \left[\dfrac{dx}{dy}\right]_{(a,0)} = 0$

$$\frac{d^2x}{dy^2} = -\frac{2}{a^3}\left[x^2 + 2xy\frac{dx}{dy}\right]$$

$$\left[\frac{d^2x}{dy^2}\right]_{(a,0)} = -\frac{2}{a^3}$$

$$\rho = \frac{[1 + \left[\dfrac{dy}{dx}\right]^2]^{\frac{3}{2}}}{\dfrac{d^2x}{dy^2}} = \frac{(1+0)^{3/2}}{-\dfrac{2}{a}} = -\frac{a}{2}$$

Therefore, $\rho = \dfrac{a}{2}$

4) Find the radius of curvature of the curve $a^3 - x^3 = xy^2$ at the point $(a,0)$.

Sol: Given: $a^3 - x^3 = xy^2$

Differentiating with respect to x,

$$2xy\frac{dy}{dx} + y^2 = -3x^2$$

$$2xy\frac{dy}{dx} = -y^2 - 3x^2$$

$$\frac{dy}{dx} = \frac{-y^2 - 3x^2}{2xy}$$

$$\left[\frac{dy}{dx}\right]_{(a,0)} = \infty$$

Consider $\dfrac{dx}{dy} = \dfrac{-2xy}{y^2 + 3x^2}$; $\left[\dfrac{dx}{dy}\right]_{(a,0)} = 0$

$$\frac{d^2x}{dy^2} = \frac{(y^2 + 3x^2)\left[2x + 2y\dfrac{dx}{dy}\right] - 2xy\left[2y + 6x\dfrac{dx}{dy}\right]}{(y^2 + 3x^2)^2}$$

$$\left[\frac{d^2x}{dy^2}\right]_{(a,0)} = -\frac{2}{3a}$$

$$\rho = \left[\frac{[1 + \left[\dfrac{dx}{dy}\right]^2]^{\frac{3}{2}}}{\dfrac{d^2x}{dy^2}}\right] = \frac{(1+0)^{3/2}}{-\dfrac{2}{3a}} = -\frac{3a}{2}$$

Therefore, $\rho = \dfrac{3a}{2}$

5) Find the radius of curvature of the curve $x^3 + y^3 = 3axy$ at $(\frac{3a}{2}, \frac{3a}{2})$.

Sol: Given: $x^3 + y^3 = 3axy$ ➡ (1)

Differentiating with respect to x,

$$3x^2 + 3y^2 \frac{dy}{dx} = 3a\left[x\frac{dy}{dx} + y\right]$$

$$\frac{dy}{dx}(y^2 - ax) = ay - x^2$$

$$\frac{dy}{dx} = \frac{ay - x^2}{y^2 - ax} \quad ➡ \qquad (2)$$

$$\frac{dy}{dx}\left[\frac{3a}{2}, \frac{3a}{2}\right] = -1$$

$$\frac{d^2y}{dx^2} = \frac{(y^2 - ax)\left[a\frac{dy}{dx} - 2x\right] - (ay - x^2)\left[2y\frac{dy}{dx} - a\right]}{(y^2 - ax)^2}$$

$$\frac{d^2y}{dx^2}\left[\frac{3a}{2}, \frac{3a}{2}\right] = -\frac{32}{3a}$$

$$\rho = \frac{[1+[y_1]^2]^{\frac{3}{2}}}{y_2} = \frac{(1+1)^{\frac{3}{2}}}{-32}(3a) = -\frac{3\sqrt{2}a}{16}$$

Therefore, $\rho = \dfrac{3\sqrt{2}a}{16}$

6) Show that the radius of curvature of the hypocycloid $x^{2/3} + y^{2/3} = a^{2/3}$ at $(a\cos^3\theta, a\sin^3\theta)$ is $3\,a\sin\theta\cos\theta$.

Sol: Given: $x^{2/3} + y^{2/3} = a^{2/3}$

The parametric equations of the curve are

$$x = a\cos^3\theta \text{ and } y = a\sin^3\theta$$

$$\frac{dx}{d\theta} = -3a\cos^2\theta\sin\theta \text{ and } \frac{dy}{d\theta} = 3a\sin^2\theta\cos\theta$$

$$\frac{dy}{dx} = -\frac{3a\sin^2\theta\cos\theta}{3a\cos^2\theta\sin\theta} = -\tan\theta$$

$$\frac{d^2y}{dx^2} = -\sec^2\theta\frac{d\theta}{dx} = \frac{-\sec^2\theta}{-3a\cos^2\theta\sin\theta} = \frac{1}{3a\cos^4\theta\sin\theta}$$

$$\rho = \frac{[1+[y_1]^2]^{\frac{3}{2}}}{y_2} = \left[1+\tan^2\theta\right]^{\frac{3}{2}} \cdot 3a\cos^4\theta\sin\theta = 3a\sin\theta\cos\theta$$

therefore, $\rho = 3a\sin\theta\cos\theta$.

7) Show that for the curve $y = \dfrac{ax}{a+x}$, the radius of curve ρ at (x,y) is related as $\left(\dfrac{2\rho}{a}\right)2/3 = \dfrac{x^2}{y^2}+\dfrac{y^2}{x^2}$.

Sol: Given: $y = \dfrac{ax}{a+x}$ ➜ $\dfrac{y}{x} = \dfrac{a}{a+x}$ ➜ (1)

Differentiating the given equation with respect to 'x', we get

$$y_1 = \frac{(a+x)a - ax(1)}{(a+x)^2} = a^2(a+x)^{-2} ➜ \qquad (2)$$

Again, differentiating (2) with respect to 'x', we get

$$y_2 = (-2)a^2(a+x)^{-3} ➜ \qquad (3)$$

Using (1), (2) and (3), we get $y_1 = \dfrac{y^2}{x^2}$ and $y_2 = -\dfrac{2}{a}\dfrac{y^3}{x^3}$

So, $\rho = \dfrac{\left[1+\dfrac{y^4}{x^4}\right]^{\frac{3}{2}}}{-\dfrac{2}{a}\dfrac{y^3}{x^3}}$

$$\rho = \dfrac{\left[1+\dfrac{y^4}{x^4}\right]^{\frac{3}{2}}\left[\dfrac{x^3}{y^3}\right]}{-\dfrac{2}{a}}$$

$$\rho = -\dfrac{a}{2}\left[\left[1+\dfrac{y^4}{x^4}\right]\left[\dfrac{x^3}{y^3}\right]^{\frac{2}{3}}\right]^{\frac{3}{2}}$$

$$\rho = -\dfrac{a}{2}\left[\left[1+\dfrac{y^4}{x^4}\right]\dfrac{x^2}{y^2}\right]^{\frac{3}{2}}$$

Therefore, $\left(\dfrac{2\rho}{a}\right)2/3 = \dfrac{x^2}{y^2}+\dfrac{y^2}{x^2}$

PARTIAL DIFFERENTIAL EQUATIONS

2 MARKS

1) Find the nth derivative of $\sin^3 2x$

Sol: Let $y = \sin^3 2x$

$$= \frac{3}{4} \sin 2x - \frac{1}{4} \sin 6x \qquad (\sin 3A = 3\sin A - 4\sin^3 A)$$

Therefore, $y_n = \frac{3}{4} . 2^n \sin\left[2x + \frac{n\pi}{2}\right] - \frac{1}{4} . 6^n \sin\left[6x + \frac{n\pi}{2}\right].$

2) Write the pedal equation of the circle $x^2 + y^2 = a^2$.

Sol: In this case of a circle, the perpendicular distance from the centre on the tangent at any point is equal to the radius. (i.e.,) p = r. This is the p-r equation of the circle.

3) Find n^{th} derivative of $\sin(ax+b)$.

Sol: Let $y = \sin(ax+b)$

$$y_1 = \cos(ax + b). a = a \sin\left[ax + b + \frac{\pi}{2}\right]$$

$$y_2 = a \cos\left[ax + b + \frac{\pi}{2}\right].a = a^2 \sin\left[ax + b + 2\frac{\pi}{2}\right]$$

Differentiating 'n' times, $y_n = a^n \sin\left[ax + b + n\frac{\pi}{2}\right]$

4) Form the P.D.E by eliminating the arbitrary function in $z = f(x^2 + y^2)$.

Sol: Given: $z = f(x^2 + y^2)$ ➜ (1)

Differentiating (1) partially with respect to x, we get

$p = f^1(x^2 + y^2)2x$ ➜ (2)

Differentiating (1) partially with respect to y, we get

$q = f^1(x^2 + y^2)2y$ ➜ (3)

$$\frac{(2)}{(3)} = \frac{p}{q} = \frac{2x}{2y}$$

Therefore $yp - xq = 0$ is the required PDE.

5) Solve $\sqrt{p} + \sqrt{q} = 1$.

Sol: Given: $\sqrt{p} + \sqrt{q} = 1$ ➜ (1)

The given equation is of the form $F(p, q) = 0$

Let $z = ax + by + c$ be a solution of (1)

Therefore, $p = a$ and $q = b$

Therefore, $\sqrt{a} + \sqrt{b} = 1$

$$\sqrt{b} = 1 - \sqrt{a} \quad ➜ \quad b = \left(1 - \sqrt{a}\right)^2$$

Hence the complete integral is

$$z = ax + \left(1 - \sqrt{a}\right)^2 y + c \quad ➜ \qquad (2)$$

To get singular integral, differentiating (2) partially with respect to a and c and equating to zero, we get

$$0 = x + 2\left(1 - \sqrt{a}\right) - \frac{-1}{2\sqrt{a}}$$

0 = 1 which is absurd.

Hence there is no singular integral.

6) Form the P.D.E by eliminating the arbitrary function in z = f(x - y).

Sol: Given: z = f(x - y) ➜ (1)

Partially differentiating (1) with respect to 'x', we get

$$p = \frac{\partial z}{\partial x} = f^{|}\left(x + y\right)\frac{\partial}{\partial x}(x + y) = f^{|}(x + y) \quad ➜ \quad (2)$$

Partially differentiating (1) with respect to 'y', we get

$$q = \frac{\partial z}{\partial y} = f^{|}\left(x + y\right)\frac{\partial}{\partial y}(x + y) = f^{|}(x + y) \quad ➜ \quad (3)$$

(3) – (2) ➜ q – p = 0 ➜ p = q.

7) Solve p + q = pq.

Sol: Given: p + q = pq ➜ (1)

The given equation is of the form F(p, q) = 0

Let z = ax + by + c be a solution of (1)

Therefore, p = a and q = b

    (1) ➜ a + b = ab

    ➜ b(1 – a) = – a

    ➜ $b = \dfrac{a}{a-1}$

Therefore, $z = ax + \dfrac{a}{a-1}y + c$ ➜ (2)

Which is the complete integral of (1), since it contains arbitrary constants. To get the singular integral, differentiating (2) partially with respect to a and c and equating them to zero, we get

$$0 = x + \frac{(a-1).1-(a+1)}{(1-a)^2} = x - \frac{1}{(1-a)^2}$$

$0 = 1$ which is absurd.

Therefore, there is no singular integral.

8)   Eliminate the arbitrary function f from $f(x^2 + y^2, z - xy) = 0$.

Sol: Given: $f(x^2 + y^2, z - xy) = 0$

Denote $x^2 + y^2$ by u and $z - xy$ by v. Then the eliminant of f is the P.D.E.,

$$\begin{vmatrix} p & q & -1 \\ u_x & u_y & u_z \\ v_x & v_y & v_z \end{vmatrix} = 0 \quad \Rightarrow \quad \begin{vmatrix} p & q & -1 \\ 2x & 2y & 0 \\ -y & -x & 1 \end{vmatrix} = 0$$

i.e., $p(2y) - q(2x) - (-2x^2 + 2y^2) = 0$

$py - qx + x^2 - y^2 = 0$, which is the required equation.

9)   Solve pq = 1.

Sol: Given: pq = 1 $\Rightarrow$ (1)

The given equation is of the form F(p, q) = 0

Let z = ax + by + c be a solution of   (1)

Therefore, p = a and q = b

     (1) $\Rightarrow$ ab = 1

     $\Rightarrow b = \dfrac{1}{a}$

Hence $z = ax + \dfrac{1}{a}y + c$ is the complete integral since it contains two arbitrary constants.

5 MARKS

1) If $y = e^{mx} \sin(ax + b)$, find y_n.

Sol: Given: $y = e^{mx} \sin(ax + b)$

$$y_1 = (me^{mx}) \sin(ax + b) + e^{mx} [a \cos(ax + b)]$$

$$= e^{mx} [m \sin(ax + b) + a \cos(ax + b)]$$

Let $m = r \cos\alpha$ and $a = r \sin\alpha$. Then $r = \sqrt{m^2 + a^2}$, $\alpha = \tan^{-1}(a/m)$

$$y_1 = e^{mx} r [\sin(ax + b)\cos\alpha + \cos(ax + b)\sin\alpha]$$

$$= e^{mx} r \sin(ax + b + \alpha)$$

So, the effect of differentiating y once is to multiply $e^{mx} \sin(ax + b)$ by r and then to increase the angle $ax + b$ by α. Thus $y_n = e^{mx} r^n \sin(ax + b + n\alpha)$.

2) Find the radius of curvature at any point on the curve $x = e^t \cos t$, $y = \sin t$.

Sol: Given: $x = e^t \cos t$, $y = \sin t$

$$\frac{dx}{dt} = e^t(\cos t - \sin t), \frac{dy}{dt} = e^t(\cos t + \sin t)$$

$$y_1 = \frac{dy}{dx} = \frac{\cos t + \sin t}{\cos t - \sin t} = \frac{1 + \tan t}{1 - \tan t} = \tan\left(\frac{\pi}{4} + t\right)$$

$$y_2 = \frac{d^2 y}{dx^2} = \sec^2\left(\frac{\pi}{4} + t\right) \cdot \frac{1}{e^t(\cos t - \sin t)}$$

$$\rho = \frac{\left[1 + \tan^2\left(\dfrac{\pi}{4} + t\right)\right]^{\frac{3}{2}}}{\sec^2\left(\dfrac{\pi}{4} + t\right)} \cdot e^t(\cos t - \sin t) = \frac{\left[\sec^2\left(\dfrac{\pi}{4} + t\right)\right]^{\frac{3}{2}} e^t(\cos t - \sin t)}{\sec^2\left(\dfrac{\pi}{4} + t\right)}$$

$$= \frac{sec^2\left(\dfrac{\pi}{4}+t\right).e^t\,(\cos t - \sin t)}{sec^2\left(\dfrac{\pi}{4}+t\right)} \qquad = \frac{e^t\,(\cos t - \sin t)}{\cos\left(\dfrac{\pi}{4}+t\right)}$$

$$= \frac{e^t\,(\cos t - \sin t)}{\cos\dfrac{\pi}{4}\cos t - \sin\dfrac{\pi}{4}\sin t} \qquad = \sqrt{2}\,e^t.$$

3) Show that in the equiangular spiral r = $ae^{\theta \cot \alpha}$ the tangent is inclined at constant angle to the radius vector.

Sol: The equation of the equiangular spiral is r = $ae^{\theta \cot \alpha}$

Taking log on both sides, log r = log a + $\theta \cot \alpha$

Differentiating with respect to θ, $\dfrac{1}{r}\dfrac{dr}{d\theta} = \cot\alpha$

$$r\frac{d\theta}{dr} = \frac{1}{\cot\alpha} = \tan\alpha$$

i.e., $\tan\phi = \tan\alpha$ ➜ $\phi = \alpha$ which is a constant.

Hence the tangent makes a constant angle with the radius vector.

4) Show that the radius of curvature at the point (x,y) on the curve y = c cosh x/c is $\dfrac{y^2}{c}$.

Sol: Given: y = c cosh x/c

$$\frac{dy}{dx} = c.\sinh\frac{x}{c}.\frac{1}{c} = \sinh\frac{x}{c}$$

$$\frac{d^2 y}{dx^2} = \frac{1}{c}\cosh\frac{x}{c}$$

$$\rho = \frac{[1+[y_1]^2]^{\frac{3}{2}}}{y_2} = \frac{\left[1+\sinh^2\frac{x}{c}\right]^{\frac{3}{2}}}{\frac{1}{c}\cosh\frac{x}{c}} = \frac{\cosh^3\frac{x}{c}.c}{\cosh\frac{x}{c}} = c\cosh^2\frac{x}{c} = c.\frac{y^2}{c^2} = \frac{y^2}{c}.$$

5) Find the slope of the tangent to the curve r = a (1 – cos θ) at θ = π/2.

Sol: Given: r = a (1 – cos θ)

$$\frac{dr}{d\theta} = a\sin\theta$$

$$tan\phi = r\frac{d\theta}{dr} = \frac{a(1-\cos\theta)}{a\sin\theta}$$

$$= \frac{2\sin^2\frac{\theta}{2}}{2\sin\frac{\theta}{2}\cos\frac{\theta}{2}}$$

$$\phi = \frac{\theta}{2}$$

With the usual notations, $\varphi = \theta + \phi = \theta + \frac{\theta}{2} = \frac{3\theta}{2}$

At $\theta = \frac{\pi}{2}$, $\varphi = \frac{3\pi}{4}$

Slope of the tangent $= \tan\varphi = \tan\frac{3\pi}{4} = -1.$

6) If $y = (x+\sqrt{1+x^2})^m$, then prove that $(1+x^2)y_{n+2} + (2n+1)xy_{n+1} + (n^2 - m^2)y_n = 0.$

Sol: Given: $y = (x+\sqrt{1+x^2})^m$

Differentiating with respect to x, we get

$$\frac{dy}{dx} = m\left[x+\sqrt{1+x^2}\right]^{m-1}\left[1+\frac{2x}{2\sqrt{1+x^2}}\right]$$

$$\frac{dy}{dx} = \frac{m\left[x+\sqrt{1+x^2}\right]^{m-1}\left[\sqrt{1+x^2}+x\right]}{\sqrt{1+x^2}}$$

$$\frac{dy}{dx} = \frac{m\left[x+\sqrt{1+x^2}\right]^{m}}{\sqrt{1+x^2}}$$

$$\frac{dy}{dx} = \frac{my}{\sqrt{1+x^2}}$$

Cross multiplying and squaring on both sides, we get

$$(1+x^2)\left[\frac{dy}{dx}\right]^2 = m^2 y^2$$

Again differentiating with respect to x, we get

$$(1+x^2)2\frac{dy}{dx}\cdot\frac{d^2 y}{dx^2} + \left[\frac{dy}{dx}\right]^2 .2x = m^2 .2y.\frac{dy}{dx}$$

Cancelling $2\frac{dy}{dx}$, we get

$$(1+x^2)\frac{d^2 y}{dx^2} + x\frac{dy}{dx} - m^2 y = 0 \quad \Rightarrow \quad (1+x^2)\, y_2 + xy_1 - m^2 y = 0$$

Differentiating 'n' times, we get $(1+x^2)y_{n+2} + (2n+1)xy_{n+1} + (n^2 - m^2)y_n = 0$.

7) Find the angle between the radius vector and the tangent at any point for the curve $\dfrac{l}{r} = (1+e\cos\theta)$.

Sol: Given: $\dfrac{l}{r} = (1+e\cos\theta)$.

Differentiating with respect to r, we get

$$-\frac{\iota}{r^2}\frac{dr}{d\theta} = -e\sin\theta$$

$$\frac{dr}{d\theta} = e\left[\frac{r^2}{\iota}\right]\sin\theta$$

$$tan\,\phi = r\frac{d\theta}{dr} = r\left[\frac{1}{er^2\sin\theta}\right] = \frac{1}{er\sin\theta}$$

$$tan\,\phi = \frac{1+e\cos\theta}{e\sin\theta}$$

$$\phi = \tan^{-1}\left[\frac{1+e\cos\theta}{e\sin\theta}\right].$$

8) Solve $z = px + qy + p^2q^2$.

Sol: Given: $z = px + qy + p2q2$ ➜   (1)

The complete integral of (1) is $z = ax + by + a^2b^2$ ➜   (2)

To get the singular integral, find $\dfrac{\partial z}{\partial a}$ and $\dfrac{\partial z}{\partial b}$ equate them to zero.

    $x + 2ab^2 = 0$ ➜   (3)

    $y + 2a^2b = 0$ ➜   (4)

From (3) and (4), $a = \dfrac{-x}{2b^2}$ and $b = \dfrac{-y}{2a^2}$

$$b = \left[\frac{-x^2}{2y}\right]^{\frac{1}{3}} \;➜\; (5) \quad \alpha\nu\delta \quad a = \left[\frac{-y^2}{2x}\right]^{\frac{1}{3}} \;➜\; (6)$$

Putting (5) and (6) in (2), we get the simple integral.

9) Solve $p + q = \sin x + \sin y$.

Sol: Given: $p + q = \sin x + \sin y$

Firstly, we'll rearrange the given equation by shifting q to the right-hand side and sin x to the left-hand side of the given equation, we get

$$p - \sin x = \sin y - q$$

Now, allow us assume $a = p - \sin x = \sin y - q$

Further, we'll simplify the worth of p and q from above, we get

$p = a + \sin x$

$q = \sin y - a$

Furthermore, we'll find $dz = pdx + qdy$ by substituting the values of p and q, we get

$dz = (a + \sin x)\,dx + (\sin y - a)\,dy$

Now, we'll integrate the above equation, we get

$$\int dz = \int (a + \sin x)dx + \int (\sin y - a)dy$$
$$z = ax - \cos x - \cos y - ay + C$$

The above equation can not be integrated more.

Hence, the entire solution of

$p + q = \sin x + \sin y$ is $z = ax - \cos x - \cos y - ay + \text{C}.$

10) Solve $z = px + qy + 2\sqrt{pq}$.

Sol: Given: $z = px + qy + 2\sqrt{pq}$ ➜ (1)

The complete integral of (1) is $z = ax + by + 2\sqrt{ab}$ ➜ (2)

To get the singular integral, find $\dfrac{\partial z}{\partial a}$ and $\dfrac{\partial z}{\partial b}$ equate them to zero.

$$x + \sqrt{\frac{b}{a}} = 0 \;➜\; (3) \text{ and } y + \sqrt{\frac{a}{b}} = 0 \;➜\; (4)$$

From (3) and (4), $xy = 1$, which is the S.I.

11) Solve $p - x^2 = q + y^2$.

Sol: Given: $p - x^2 = q + y^2$

Let $p - x^2 = q + y^2 = k$

$$p - x^2 = k \text{ and } q + y^2 = k$$

$$\rightarrow p = x^2 + k \text{ and } q = k - y^2$$

Now, dz = pdx + qdy

$$dz = (x^2 + k)dx + (k - y^2)dy$$

Therefore, $z = \dfrac{x^3}{3} + kx + ky - \dfrac{y^3}{3} + b$

$$z = \frac{x^3}{3} - \frac{y^3}{3} + (x + y)k + b \text{ which is the C.I.}$$

12) Eliminate f and g from z = f(x + ay) + g(x + by).

Sol: Given: z = f(x + ay) + g(x + by)

From the given equation, we get

$$p = \frac{\partial z}{\partial x} = f'(x + ay) + g'(x + by)$$

$$q = \frac{\partial z}{\partial y} = f'(x + ay)a + g'(x + by)b$$

$$r = \frac{\partial^2 z}{\partial x^2} = f''(x + ay) + g''(x + by)$$

$$s = \frac{\partial^2 z}{\partial x \partial y} = f''(x + ay)a + g''(x + by)b$$

$$t = \frac{\partial^2 z}{\partial y^2} = f''(x + ay)a^2 + g''(x + by)b^2$$

Therefore, ar − s = (a − b)g''(x+by) and as − t = b(a − b)g''(x+by)

From the above equations, we have ab(r+t) = (a+b)s.

13) Solve $q = px + p^2$.

Sol: Let $q = px + p^2 = \dfrac{k^2}{4}$

Therefore, $q = \dfrac{k^2}{4}$ and $px + p^2 = \dfrac{k^2}{4} = \dfrac{-x \pm \sqrt{x^2 + k^2}}{2}$

Now $dz = pdx + qdy$

$$dz = \left[\frac{-x}{2} \pm \frac{\sqrt{x^2 + k^2}}{2} \right] dx + \frac{k^2}{4} dy$$

Therefore, $z = \dfrac{-x^2}{4} \pm \dfrac{1}{2}\left[\dfrac{k^2}{2} \sinh^{-1} \dfrac{x}{k} + \dfrac{x\sqrt{x^2 + k^2}}{2} \right] + \dfrac{k^2}{4} y + b$

Which is the complete integral.

10 MARKS

1) Solve $(mz - ny)p + (nx - lz)q = ly - mx$.

Sol: Given: $(mz - ny)p + (nx - lz)q = ly - mx$

It is of the form $Pp + Qq = R$

$P = mz - ny,\ Q = nx - lz,\ R = ly - mx.$

Subsidiary equations are $\dfrac{dx}{mz - ny} = \dfrac{dy}{nx - lz} = \dfrac{dz}{ly - mx}$ → (1)

Using the multipliers x, y, z each of the rations in (1)

$$= \frac{xdx + ydy + zdz}{x(mz - ny) + y(nx - lz) + z(ly - mx)}$$

$$= \frac{xdx + ydy + zdz}{0}$$

$xdx + ydy + zdz = 0$

Integrating, we get $x^2 + y^2 + z^2 = a$

Using the multipliers l, m, n each of the ratios in (1)

$$= \frac{ldx + mdy + ndz}{l(mz - ny) + m(nx - lz) + n(ly - mx)}$$

$$= \frac{ldx + mdy + ndz}{0}$$

$$= ldx + mdy + ndz = 0$$

Integrating, we get lx + my + nz = b

Hence the solution is $\phi(x^2 + y^2 + z^2, lx + my + nz) = 0$.

2)   Solve $x(z^2 - y^2)p + y(x^2 - z^2)q = z(y^2 - x^2)$,

Sol: Given: $x(z^2 - y^2)p + y(x^2 - z^2)q = z(y^2 - x^2)$

It is of the form Pp + Qq = R

$P = x(z^2 - y^2)$, $Q = y(x^2 - z^2)$, $R = z(y^2 - x^2)$

Subsidiary equations are $\dfrac{dx}{x(z^2 - y^2)} = \dfrac{dy}{y(x^2 - z^2)} = \dfrac{dz}{z(y^2 - x^2)}$ ➔   (1)

Using the multipliers x, y, z each of the term in (1)

$$= \frac{xdx + ydy + zdz}{x^2(z^2 - y^2) + y^2(x^2 - z^2) + z^2(y^2 - x^2)}$$

$$= \frac{xdx + ydy + zdz}{0}$$

xdx + ydy + zdz = 0

Integrating, we get $x^2 + y^2 + z^2 = a$

Now, $\dfrac{\frac{dx}{x}}{(z^2 - y^2)} = \dfrac{\frac{dy}{y}}{(x^2 - z^2)} = \dfrac{\frac{dz}{z}}{(y^2 - x^2)}$ ➔   (2)

Using the multipliers 1, 1, 1 each of the term in (2)

$$= \frac{\dfrac{dx}{x} + \dfrac{dy}{y} + \dfrac{dz}{z}}{z^2 - y^2 + x^2 - z^2 + y^2 - x^2}$$

$$= \frac{\dfrac{dx}{x} + \dfrac{dy}{y} + \dfrac{dz}{z}}{0}$$

$$\frac{dx}{x} + \frac{dy}{y} + \frac{dz}{z} = 0$$

Integrating, we get log x + log y + log z = log b ➜ xyz = b

Hence the solution is $\phi(x^2 + y^2 + z^2, xyz) = 0$

3) Solve x(y − z)p + y(z − x)q = z(z − y).

Sol: Given: x(y − z)p + y(z − x)q = z(z − y)

It is of the form Pp + Qq = R

P = x(y − z), Q = y(z − x), R = z(x − y)

Subsidiary equations are $\dfrac{dx}{x(y-z)} = \dfrac{dy}{y(z-x)} = \dfrac{dz}{z(x-y)}$ ➜ (1)

Using the multipliers 1, 1, 1 each of the term in (1)

$$= \frac{dx + dy + dz}{x(y-z) + y(z-x) + z(x-y)}$$

$$= \frac{dx + dy + dz}{0}$$

dx + dy + dz = 0

Integrating, we get x + y + z = a.

Now, $\dfrac{\dfrac{dx}{x}}{(y-z)} = \dfrac{\dfrac{dy}{y}}{(z-x)} = \dfrac{\dfrac{dz}{z}}{(x-y)}$ ➜ (2)

Using the multipliers 1, 1, 1 each of the term in (2)

$$= \frac{\dfrac{dx}{x} + \dfrac{dy}{y} + \dfrac{dz}{z}}{y-z+z-x+x-y}$$

$$= \frac{dx}{x} + \frac{dy}{y} + \frac{dz}{z} = 0$$

Integrating, we get log x + log y + log z = log b ➜ xyz = b

Hence the solution is $\phi(x + y + z, xyz) = 0$.

ALLIED MATHEMATICS – II

INTEGRAL CALCULUS

2 MARKS

1) Evaluate $\int x^4 \cos x\, dx$.

Sol: Let $I = \int x^4 \cos x\, dx$

We know that $\int = uv_1 - u^1 v_2 + u^2 v_3 - \dots$

$$v = \cos x \qquad\qquad u = x^4$$

$$v_1 = \sin x \qquad\qquad u^1 = 4x^3$$

$$v_2 = -\cos x \qquad\qquad u^2 = 12x^2$$

$$v_3 = -\sin x \qquad\qquad u^3 = 24x$$

$$v_4 = \cos x \qquad\qquad u^4 = 24$$

$$v_5 = \sin x \qquad\qquad u^5 = 0$$

Now, $I = uv_1 - u^1 v_2 + u^2 v_3 - \dots$

$$= x^4[\sin x] - 4x^3[-\cos x] + 12x^2[-\sin x] - 24x[\cos x] + 24[\sin x]$$

$$= \sin x\,[x^4 - 12x^2 + 24] + \cos x\,[4x^3 - 24x].$$

2) Evaluate $\int_0^{\frac{\pi}{2}} \sin^6 x\, dx$.

Sol: We know that $I_n = \dfrac{n-1}{n} \cdot \dfrac{n-3}{n-2} \cdots \dfrac{2}{3}$ when n is odd.

Therefore, $I_6 = \dfrac{5}{6} \cdot \dfrac{3}{4} \cdot \dfrac{1}{2} \cdot \dfrac{\pi}{2} = \dfrac{5\pi}{32}.$

3) Evaluate $\int x^2 e^{-2x}\, dx$.

Sol: Let $I = \int x^2 e^{-2x}\, dx$

We know that $= uv_1 - u^1 v_2 + u^2 v_3 - \ldots$

$$v = e^{-2x} \qquad u = x^2$$

$$v_1 = \frac{e^{-2x}}{-2} \qquad u^1 = 2x$$

$$v_2 = \frac{e^{-2x}}{4} \qquad u^2 = 2$$

$$v_3 = \frac{e^{-2x}}{-8} \qquad u^3 = 0$$

Now, $I = uv_1 - u^1 v_2 + u^2 v_3 - \ldots$

$$= x^2\left[\frac{e^{-2x}}{-2}\right] - 2x\left[\frac{e^{-2x}}{4}\right] + 2\left[\frac{e^{-2x}}{-8}\right] + c$$

$$= -\frac{e^{-2x}}{4}\left[2x^2 + 2x + 1\right] + c.$$

4) Evaluate $\int_0^{\frac{\pi}{2}} \sin^6\theta \cos^4\theta\, d\theta$.

Sol: Using the reduction formula for Im,n we have

$$\int_0^{\frac{\pi}{2}} \sin^6 è \cos^4 è\, dè = \frac{3}{10}\cdot\frac{1}{8}\cdot\frac{5}{6}\cdot\frac{3}{4}\cdot\frac{1}{2}\cdot\frac{\pi}{2} = \frac{3\pi}{512}.$$

5) Evaluate $\int x^4 \sin x\, dx$.

Sol: Let $I = \int x^4 \sin x\, dx$

We know that $= uv_1 - u^1 v_2 + u^2 v_3 - \ldots$

$$v = \sin x \qquad u = x^4$$

$$v_1 = -\cos x \qquad u^1 = 4x^3$$

$$v_2 = -\sin x \qquad u^2 = 12x^2$$

$$v_3 = \cos x \qquad\qquad u^3 = 24x$$

$$v_4 = \sin x \qquad\qquad u^4 = 24$$

$$v_5 = -\cos x \qquad\qquad u^5 = 0$$

Now, $I = uv_1 - u^1v_2 + u^2v_3 - \ldots$

$$= x^4[-\cos x] - 4x^3[-\sin x] + 12x^2[\cos x] - 24x[\sin x] + 24[-\cos x]$$

$$= \cos x\,[-x^4 + 12x^2 - 24] + \sin x\,[4x^3 - 24x].$$

6) Evaluate $\int_0^{\frac{\pi}{2}} x^2 \sin x\, dx$.

Sol: Let $I = \int_0^{\frac{\pi}{2}} x^2 \sin x\, dx$.

We know that $= uv_1 - u^1v_2 + u^2v_3 - \ldots$

$$v = \sin x \qquad\qquad u = x^2$$

$$v_1 = -\cos x \qquad\qquad u^1 = 2x$$

$$v_2 = -\sin x \qquad\qquad u^2 = 2$$

$$v_3 = \cos x \qquad\qquad u^3 = 0$$

Now, $I = uv_1 - u^1v_2 + u^2v_3 - \ldots$

$$= \left[-x^2\cos x + 2x\sin x + 2\cos x\right]_0^{\frac{\pi}{2}} = \pi - 2.$$

7) Evaluate $\int x \log x\, dx$.

Sol: Let $I = \int x \log x\, dx$

We know that $\int u\, dv = uv - \int v\, du$

$$u = \log x \qquad dv = x\, dx$$

$$du = \frac{1}{x}dx \quad v = \frac{x^2}{2}$$

Now, $I = \dfrac{x^2}{2}\log x - \int \dfrac{x^2}{2}\dfrac{1}{x}\, dx$

$$\frac{x^2}{2}\log x - \frac{x^2}{4}$$

8) Evaluate $\int_0^{\frac{\pi}{2}} \sin^7\theta \, d\theta$.

Sol: We know that $I_n = \frac{n-1}{n}\cdot\frac{n-3}{n-2}\ldots\frac{2}{3}$ when n is odd.

Therefore, $I_7 = \frac{6}{7}\cdot\frac{4}{5}\cdot\frac{2}{3} = \frac{48}{105}$.

5 MARKS

1) Evaluate $\int_0^{\frac{\pi}{2}} \log \sin x \, dx$.

Sol: Let $I = \int_0^{\frac{\pi}{2}} \log \sin x \, dx$ ➜   (1)

Replacing x by $\frac{\pi}{2} - x$, we get

$$I = \int_0^{\frac{\pi}{2}} \log \sin\left[\frac{\pi}{2} - x\right] dx = \int_0^{\frac{\pi}{2}} \log \cos x \, dx \ \text{➜} \ (2)$$

(1) +(2) gives $2I = \int_0^{\frac{\pi}{2}} \log(\sin x \cos x)\, dx = \int_0^{\frac{\pi}{2}} \log \sin\frac{2x}{2}\, dx$

$$= \int_0^{\frac{\pi}{2}} \log \sin 2x \, dx - \int_0^{\frac{\pi}{2}} \log 2 \, dx$$

$= I_1 - I_2$ (say)   ➜   (3)

Now, $I_2 = \frac{\pi}{2}\left[\log 2\right]$

$$I_1 = \int_0^{\frac{\pi}{2}} \log \sin 2x \, dx$$

Put $2x = t$ ➜ $2dx = dt$.

When x = 0, t = 0 and when $x = \dfrac{\pi}{2}, t = \pi$.

We know that $\displaystyle\int_0^a f(x)\,dx = 2\int_0^{\frac{a}{2}} f(x)\,dx$ if f(a − x) = f(x).

Now, log sin(π − t) = log sin t.

Therefore, $I_1 = \dfrac{1}{2}\left[2\int_0^{\frac{\pi}{2}} \log \sin 2x\,dx \right] = I$

$$(3) \;\blacktriangleright\; 2I = I - \dfrac{\pi}{2}\big[\log 2\big] \;\blacktriangleright\; I = \dfrac{\pi}{2}\left[\log \dfrac{1}{2}\right]$$

2) Show that $\displaystyle\int_0^\pi x \sin^6 x\,dx = \dfrac{5\pi^2}{32}$.

Sol: Let $I = \displaystyle\int_0^\pi x \sin^6 x\,dx$

Replacing x by in the integrand, we get

$$I = \int_0^\pi (\pi - x)\sin^6(\pi - x)\,dx = \int_0^\pi (\pi - x)\sin^6 x\,dx = \int_0^\pi \pi \sin^6 x\,dx - I$$

Therefore, $2I = \pi\displaystyle\int_0^\pi \sin^6 x\,dx \to I = \dfrac{\pi}{2}\int_0^\pi \sin^6 x\,dx$

We know that if f(a − x) = f(x), then $\displaystyle\int_0^a f(x)\,dx = 2\int_0^{\frac{a}{2}} f(x)\,dx$

Now, $\sin^6 (\pi - x) = \sin^6 x$.

Therefore, $I = \dfrac{\pi}{2}\left[2\int_0^{\frac{\pi}{2}} \sin^6 x\,dx \right] = \pi I_6 = \pi\left[\dfrac{5}{6}\cdot\dfrac{3}{4}\cdot\dfrac{1}{2}\cdot\dfrac{\pi}{2}\right] = \dfrac{5\pi^2}{32}$.

3) Evaluate $\displaystyle\int x^4 \sin\left[\dfrac{x}{2}\right] dx$.

Sol: Let $I = \displaystyle\int x^4 \sin\left[\dfrac{x}{2}\right] dx$

We know that $\displaystyle\int uv\,dx = uv_1 - u^1 v_2 + u^2 v_3 - \dots$

$$v = \sin\dfrac{x}{2} \qquad\qquad u = x^4$$

$$v_1 = -2\cos\dfrac{x}{2} \qquad\qquad u^1 = 4x^3$$

$$v_2 = -4\sin\frac{x}{2} \qquad\qquad u^2 = 12x^2$$

$$v_3 = 8\cos\frac{x}{2} \qquad\qquad u^3 = 24x$$

$$v_4 = 16\sin\frac{x}{2} \qquad\qquad u^4 = 24$$

$$v_5 = -32\cos\frac{x}{2} \qquad\qquad u^5 = 0$$

Now, $I = uv_1 - u^1 v_2 + u^2 v_3 - \ldots$

$$= -2x^4\left[\cos\frac{x}{2}\right] + 16x^3\left[\sin\frac{x}{2}\right] + 96x^2\left[\cos\frac{x}{2}\right] - 384x\left[\sin\frac{x}{2}\right] - 768\left[\cos\frac{x}{2}\right] + c.$$

4)   Evaluate $\displaystyle\int_0^{\frac{\pi}{2}} \frac{\sqrt{\sin x}}{\sqrt{\sin x} + \sqrt{\cos x}}\, dx.$

Sol: Let $\displaystyle I = \int_0^{\frac{\pi}{2}} \frac{\sqrt{\sin x}}{\sqrt{\sin x} + \sqrt{\cos x}}\, dx$ ➜ (1)

$$= \int_0^{\frac{\pi}{2}} \frac{\sqrt{\sin\left(\frac{\pi}{2} - x\right)}}{\sqrt{\sin\left(\frac{\pi}{2} - x\right)} + \sqrt{\cos\left(\frac{\pi}{2} - x\right)}}\, dx$$

$$I = \int_0^{\frac{\pi}{2}} \frac{\sqrt{\cos x}}{\sqrt{\sin x} + \sqrt{\cos x}}\, dx \quad ➜ \qquad (2)$$

Adding (1) and (2), $\displaystyle 2I = \int_0^{\frac{\pi}{2}} \frac{\sqrt{\sin x} + \sqrt{\cos x}}{\sqrt{\sin x} + \sqrt{\cos x}}\, dx$

$$= \int_0^{\frac{\pi}{2}} dx \qquad = \frac{\pi}{2}$$

Therefore, $I = \dfrac{\pi}{4}.$

5)   Show that $I = \dfrac{3\pi^2}{16}$, where $\displaystyle I = \int_0^{\pi} x\cos^4 x\, dx.$

Sol: Let $\displaystyle I = \int_0^{\pi} x\cos^4 x\, dx$

Replacing x by $\pi - x$ in the integrand, we get

$$I = \int_0^\pi (\pi - x)\cos^4(\pi - x)\, dx = \int_0^\pi (\pi - x)\cos^4 x\, dx = \int_0^\pi \pi \cos^4 x\, dx - I$$

Therefore, $2I = \pi \int_0^\pi \cos^4 x\, dx \Rightarrow I = \dfrac{\pi}{2}\int_0^\pi \cos^4 x\, dx$

We know that if $f(a - x) = f(x)$, then $\int_0^a f(x)\, dx = 2\int_0^{\frac{a}{2}} f(x)\, dx$

Now, $\cos^4(\pi - x) = \cos^4 x$.

Therefore, $I = \dfrac{\pi}{2}\left[2\int_0^{\frac{\pi}{2}} \cos^4 x\, dx\right] = \pi I_4 = \pi\left[\dfrac{3}{4}\cdot\dfrac{1}{2}\cdot\dfrac{\pi}{2}\right] = \dfrac{3\pi^2}{16}$.

6)   Evaluate $\int x^2 \sin x\, dx$.

Sol: Let $I = \int x^2 \sin x\, dx$

We know that $\int uv\, dx = uv_1 - u^1v_2 + u^2v_3 - \ldots$

$\quad\quad v = \sin x \quad\quad\quad\quad u = x^2$

$\quad\quad v_1 = -\cos x \quad\quad\quad u^1 = 2x$

$\quad\quad v_2 = -\sin x \quad\quad\quad u^2 = 2$

$\quad\quad v_3 = \cos x \quad\quad\quad\quad u^3 = 0$

Now, $I = uv_1 - u^1v_2 + u^2v_3 - \ldots$

$\quad\quad = [-x^2 \cos x + 2x \sin x + 2\cos x]$.

7)   Evaluate $\int_0^{\frac{\pi}{2}} \dfrac{\sin^2 x}{\sin x + \cos x}\, dx$.

Sol: Let $I = \int_0^{\frac{\pi}{2}} \dfrac{\sin^2 x}{\sin x + \cos x}\, dx \Rightarrow$ (1)

Replacing x by $\left[\dfrac{\pi}{2} - x\right]$ in (1),

$$I = \int_0^{\frac{\pi}{2}} \dfrac{\sin^2\left[\dfrac{\pi}{2} - x\right]}{\sin\left[\dfrac{\pi}{2} - x\right] + \cos\left[\dfrac{\pi}{2} - x\right]}\, dx = \int_0^{\frac{\pi}{2}} \dfrac{\sin^2 x}{\cos x + \sin x}\, dx \Rightarrow \quad\quad (2)$$

$$(1) \quad + (2) \;\Rightarrow\; 2I = \int_0^{\frac{\pi}{2}} \frac{1}{\sin x + \cos x}\, dx$$

The substitution $t = \tan\left(\dfrac{x}{2}\right)$ gives after simplification

$$2I = 2\int_0^1 \frac{1}{2t + 1 - t^2}\, dt \qquad = 2\int_0^1 \frac{1}{(\sqrt{2})^2 - (t-1)^2}\, dt$$

$$= 2.\frac{1}{2\sqrt{2}}\left[\log\frac{\sqrt{2}+(t-1)}{\sqrt{2}-(t-1)}\right]_0^1 = \frac{1}{\sqrt{2}}\left[0 - \log\left[\frac{\sqrt{2}-1}{\sqrt{2}+1}\right]\right]$$

$$2I = \frac{1}{\sqrt{2}}\log\left[\frac{\sqrt{2}+1}{\sqrt{2}-1}\right] \qquad = \frac{2}{\sqrt{2}}\log(\sqrt{2}+1)$$

Therefore, $I = \dfrac{1}{\sqrt{2}}\log(\sqrt{2}+1)$.

10 MARKS

---

1) Show that $\displaystyle\int_0^\pi \frac{x\tan x}{\sec x + \tan x}\, dx = \pi\left(\frac{\pi}{2}-1\right)$.

Sol: Let $\displaystyle I = \int_0^\pi \frac{x\tan x}{\sec x + \tan x}\, dx = \int_0^\pi \frac{x\sin x}{1+\sin x}\, dx$

Replacing x by $\pi - x$ in the integrand, we get

$$I = \int_0^\pi \frac{(\pi - x)\sin(\pi - x)}{1+\sin(\pi - x)}\, dx = \int_0^\pi \frac{(\pi - x)\sin x}{1+\sin x}\, dx$$

$$I = \pi\int_0^\pi \frac{\sin x}{1+\sin x}\, dx - I$$

Therefore, $\displaystyle 2I = \pi\int_0^\pi \frac{\sin x}{1+\sin x}\, dx$

$$= \pi \int_0^\pi \frac{\sin x(1-\sin x)}{(1-sin^2 x)}\, dx$$

$$= \pi \int_0^\pi \frac{\sin x - sin^2 x)}{cos^2 x}\, dx$$

$$= \pi \int_0^\pi (\sec x \tan x - tan^2 x)\, dx$$

$$= \pi \int_0^\pi (\sec x \tan x - sec^2 x + 1)\, dx$$

$$= \pi[\sec x - \tan x + x]_0^\pi$$

$$2I = \pi(\pi\text{-}2)$$

Therefore, $I = \pi(\dfrac{\pi}{2}-1)$.

2)   Evaluate $\int_0^{\frac{\pi}{2}} cos^n x\, dx$ and deduce the result when n is even and odd.

Sol: Let $I_n = \int_0^{\frac{\pi}{2}} \cos^n x\, dx$

$$= \int_0^{\frac{\pi}{2}} \cos^{n-1} x \cos x\, dx$$

Let $u = \cos^{n-1} x$ $dv = \cos x\, dx$

      $du = (n-1)\cos^{n-2} x\,(-\sin x\, dx)$ $v = \sin x$

Therefore, $I_n = \left[\cos^{n-1} x \sin x\right]_0^{\frac{\pi}{2}} + \int_0^{\frac{\pi}{2}} \sin x(n-1)\cos^{n-2} x \sin x\, dx$

$$= 0 + (n-1)\int_0^{\frac{\pi}{2}} \cos^{n-2} x \, sin^2 x\, dx$$

$$= (n-1)\int_0^{\frac{\pi}{2}} \cos^{n-2} x\,(1-cos^2 x)\, dx$$

$$I_n = (n-1)I_{n-2} - (n-1)I_n$$

$$I_n + (n-1)I_n = (n-1)I_{n-2}$$

$$(1 + n - 1)I_n = (n-1)I_{n-2}$$

Therefore, $I_n = \left[\dfrac{n-1}{n}\right]I_{n-2}$

When 'n' is even $\rightarrow$ $I_n = \dfrac{n-1}{n} \cdot \dfrac{n-3}{n-2} \cdot \dfrac{n-5}{n-4} \cdots\cdots \dfrac{1}{2}\dfrac{\pi}{2}$

When 'n' is odd $\rightarrow$ $I_n = \dfrac{n-1}{n} \cdot \dfrac{n-3}{n-2} \cdot \dfrac{n-5}{n-4} \cdots\cdots \dfrac{2}{3} \cdot$

3) Prove that $\displaystyle\int_0^{\frac{\pi}{4}} \log(1+\tan x)\,dx = \dfrac{\pi}{8}\log 2$

Sol: Let $\displaystyle I = \int_0^{\frac{\pi}{4}} \log(1+\tan x)\,dx$

Replacing x by $\dfrac{\pi}{4} - x$ in the integrand, we have

$$I = \int_0^{\frac{\pi}{4}} \log\left(1+\tan\left(\dfrac{\pi}{4} - x\right)\right)dx$$

$$= \int_0^{\frac{\pi}{4}} \log\left[1 + \dfrac{\tan\left[\dfrac{\pi}{4}\right] - \tan x}{1 + \tan\left[\dfrac{\pi}{4}\right]\tan x}\right]dx$$

$$= \int_0^{\frac{\pi}{4}} \log\left[1 + \dfrac{1-\tan x}{1+\tan x}\right]dx$$

$$\int_0^{\frac{\pi}{4}} \log\left[\dfrac{(1+\tan x) + (1-\tan x)}{1+\tan x}\right]dx$$

$$= \int_0^{\frac{\pi}{4}} \log\left[\dfrac{2}{1+\tan x}\right]dx$$

$$= \int_0^{\frac{\pi}{4}} \log[2] - \log[1 + \tan x]\, dx$$

$$= [x \log 2]_0^{\frac{\pi}{4}} - I$$

$$I = \frac{\pi}{4} \log 2 - I$$

$$2I = \frac{\pi}{4} \log 2$$

$$I = \frac{\pi}{8} \log 2.$$

APPLICATION OF INTEGRATION

2 MARKS

1) Evaluate $\int_0^a \int_x^a (x^2 + y^2)\,dy\,dx$.

Sol: Let $I = \int_0^a \int_x^a (x^2 + y^2)\,dy\,dx$

$$= \int_0^a \left[x^2 y + \frac{y^3}{3} \right]_x^a dx$$

$$= \int_0^a \left[ax^2 + \frac{a^3}{3} - \left[x^3 + \frac{x^3}{3} \right] \right] dx$$

$$= \left[a\frac{x^3}{3} + \frac{a^3 x}{3} - \frac{4x^3}{12} \right]_0^a$$

$$= \left[\frac{a^4}{3} + \frac{a^4}{3} - \frac{a^3}{3} \right]$$

$$= \frac{a^3}{3}.$$

2)   Find a_0, in the fourier series for the function $f(x) = \pi - x$, $0 < x < 2\pi$.

Sol: Given: $f(x) = \pi - x$

We know that $f(x) = \dfrac{a_0}{2} + \sum_{n=1}^{\infty} \left(a_n \cos nx + b_n \sin nx \right)$

To find: a_0 (constant term)

We know that $a_0 = \dfrac{1}{\pi} \int_0^{2\pi} f(x)\,dx = \dfrac{1}{\pi} \int_0^{2\pi} (\pi - x)\,dx$

$$= \dfrac{1}{\pi} \left[\pi x - \dfrac{x^2}{2} \right]_0^{2\pi} = 0.$$

3)   Evaluate $\int_0^a \int_0^b (x^2 + y^2)\,dx\,dy$.

Sol: Let $I = \int_0^a \int_0^b (x^2 + y^2)\,dx\,dy$.

$$\int_0^a \left[\dfrac{x^3}{3} + xy^2 \right]_0^b dy$$

$$= \int_0^a \left[\dfrac{b^3}{3} + by^2 \right] dy$$

$$= \left[y\dfrac{b^3}{3} + b\dfrac{y^3}{3} \right]_0^a$$

$$= \left[\dfrac{ab^3}{3} + \dfrac{a^3 b}{3} \right]$$

$$= \dfrac{ab}{3} \left[a^2 + b^2 \right].$$

4) Find the constant term in the Fourier Series expansion for f(x) = x sin x in $0 \leq x \leq 2\pi$

Sol: Given: f(x) = x sin x

We know that $f(x) = \dfrac{a_0}{2} + \sum_{n=1}^{\infty}(a_n \cos nx + b_n \sin nx)$

To find: a_0 (constant term)

We know that $a_0 = \dfrac{1}{\pi}\int_0^{2\pi} f(x)dx = \dfrac{1}{\pi}\int_0^{2\pi}(x \sin x)dx$

$$= \dfrac{1}{\pi}[x(-\cos x)+1(-\sin x)]_0^{2\pi} \qquad = -2.$$

5) **Evaluate $\iint$ over the region bounded by x = y = 0 and x = y = 2.**

Sol: $I = \int_0^2\int_0^2 dx\,dy$

$$= \int_0^2 [x]_0^2\,dy$$

$$= \int_0^2 2\,dy$$

$$= [2y]_0^2$$

$$= 4.$$

6) Evaluate $\int_0^1\int_0^2 xy^2\,dy\,dx$.

Sol: $I = \int_0^1\int_0^2 xy^2\,dy\,dx$

$$= \int_0^1 x\left[\dfrac{y^3}{3}\right]_0^2 dx$$

$$= \int_0^1 x\left[\dfrac{8}{3}\right]dx$$

$$= \frac{8}{3}\left[\frac{x^2}{2}\right]_0^1$$

$$= \frac{8}{3}\cdot\frac{1}{2} \qquad = \frac{4}{3}.$$

7) If $f(x) = \frac{1}{2}(\pi - x), 0 < x < 2\pi$, the find a_0.

Sol: Given: $f(x) = \frac{1}{2}[\pi - x]$

We know that $f(x) = \frac{a_0}{2} + \sum_{n=1}^{\infty}\left(a_n \cos nx + b_n \sin nx\right)$

To find: a_0 (constant term)

We know that $a_0 = \frac{1}{\pi}\int_0^{2\pi} f(x)dx$

$$= \frac{1}{\pi}\int_0^{2\pi}\frac{1}{2}(\pi - x)dx$$

$$= \frac{1}{\pi}\cdot\frac{1}{2}\left[\pi x - \frac{x^2}{2}\right]_0^{2\pi}$$

$$= 0.$$

5 MARKS

---

1) Evaluate $\int_0^1\int_0^{1-x}\int_0^{x+y} e^z \, dx \, dy \, dz$.

Sol: $I = \int_0^1\int_0^{1-x}\int_0^{x+y} e^z \, dx \, dy \, dz = \int_0^1\int_0^{1-x}\int_0^{x+y} e^z \, dz \, dy \, dx$

$$= \int_0^1\int_0^{1-x}\left[e^z\right]_0^{x+y} dy \, dx \qquad = \int_0^1\int_0^{1-x}(e^{x+y} - 1)\, dy \, dx$$

$$= \int_0^1 \int_0^{1-x} (e^x e^y - 1)\, dy\, dx = \int_0^1 \left[e^x e^y - y \right]_0^{1-x} dx$$

$$= \int_0^1 \left[e^x e^{1-x} - (1-x) - e^x \right]_0^{1-x} dx$$

$$= \int_0^1 [e - 1 + x - e^x]\, dx = \left[ex - x + \frac{x^2}{2} - e^x \right] = \frac{1}{2}.$$

2) Find the Fourier Series for the function $f(x) = \dfrac{x}{2}$ in $-\pi < x < \pi$.

Sol: Given: $f(x) = \dfrac{x}{2}$

We know that $f(x) = \dfrac{a_0}{2} + \sum_{n=1}^{\infty} (a_n \cos nx + b_n \sin nx)$

Now, $f(x)$ is an odd function. So, $a_0 = a_n = 0$.

Therefore, $f(x) = \sum_{n=1}^{\infty} (b_n \sin nx)$

$$b_n = \frac{1}{\pi} \int_{-\pi}^{\pi} f(x) \sin nx\, dx = \frac{2}{\pi} \int_0^{\pi} f(x) \sin nx\, dx$$

$$= \frac{2}{\pi} \int_0^{\pi} \frac{x}{2} \sin nx\, dx \qquad = \frac{1}{\pi} \int_0^{\pi} x \sin nx\, dx$$

$$= \frac{1}{\pi} \left[\left[-\frac{\cos nx}{n} \right](x) - \left[-\frac{\sin nx}{n^2} \right](1) \right]_0^{\pi}$$

$$= \frac{1}{\pi} \left[-\frac{\cos n\pi}{n}(\pi) \right] \qquad = -\frac{(-1)n}{n}.$$

Therefore, $b_n = \dfrac{-1}{n}$ if 'n' is even and $b_n = \dfrac{1}{n}$ if 'n' is odd.

Therefore, $\dfrac{x}{2} = \dfrac{\sin x}{1} - \dfrac{\sin 2x}{2} + \dfrac{\sin 3x}{3} - \dfrac{\sin 4x}{4} + \cdots$

3) Find the volume of the solid bounded by the surface $x = 0$, $y = 0$, $z = 0$ & $x + y + z = 1$

Sol: Given: $x = 0$, $y = 0$, $z = 0$ & $x + y + z = 1$

z varies from 0 to $1 - x - y$

y varies from 0 to $1 - x$

x varies from 0 to 1

Now, volume $= \int_0^1 \int_0^{1-x} \int_0^{1-x-y} dz\, dy\, dx$

$$= \int_0^1 \int_0^{1-x} (z)_0^{1-x-y} \, dy\, dx$$

$$= \int_0^1 \int_0^{1-x} (1-x-y) \, dy\, dx$$

$$= \int_0^1 \left[y - xy - \frac{y^2}{2} \right]_0^{1-x} dx$$

$$= \int_0^1 \left[1 - x - x(1-x) - \frac{(1-x)^2}{2} \right] dx$$

$$= \frac{1}{2} \int_0^1 (x^2 - 2x + 1) \, dx$$

$$= \frac{1}{2} \left[\frac{1}{3} - \frac{1}{2} + 1 \right]$$

$$= \frac{5}{12}.$$

4)   Find the Fourier Series for $f(x) = \dfrac{\pi - x}{2}$ in $0 < x < 2\pi$.

Sol: Given: $f(x) = \dfrac{\pi - x}{2}$

We know that $f(x) = \dfrac{a_0}{2} + \sum_{n=1}^{\infty}(a_n \cos nx + b_n \sin nx)$

Now, $f(x)$ is an odd function. So, $a_0 = a_n = 0$.

Therefore, $f(x) = \sum_{n=1}^{\infty}(b_n \sin nx)$

$$b_n = \frac{1}{\pi}\int_0^{2\pi} f(x)\sin nx\, dx = \frac{1}{\pi}\int_0^{2\pi}\left[\frac{\pi - x}{2}\right]\sin nx\, dx$$

$$= \frac{1}{2\pi}\left[\left[-\frac{\cos nx}{n}\right](\pi - x) - \left[-\frac{\sin nx}{n^2}\right](-1)\right]_0^{2\pi}$$

$$= \frac{1}{2\pi}\left[\frac{(x - \pi)\cos nx}{n} - \frac{\sin nx}{n^2}\right]_0^{2\pi} = \frac{1}{n}$$

Therefore, $f(x) = \dfrac{\sin x}{1} + \dfrac{\sin 2x}{2} + \dfrac{\sin 3x}{3} + \cdots$

5)   Find the Fourier Series for $f(x)$ in $[\pi, -\pi]$ if $f(x) = \begin{cases} 0, -\pi < x < 0 \\ \pi, 0 < x < \pi \end{cases}$

Sol: Given: $f(x) = \begin{cases} 0, -\pi < x < 0 \\ \pi, 0 < x < \pi \end{cases}$

We know that $f(x) = \dfrac{a_0}{2} + \sum_{n=1}^{\infty}(a_n \cos nx + b_n \sin nx)$

$$a_0 = \frac{1}{\pi}\left[\int_{-\pi}^0 0\, dx + \int_0^{\pi}\pi\, dx\right] = \frac{1}{\pi}\left[0 + \pi^2\right] = \pi$$

$$a_n = \frac{1}{\pi}\left[\int_{-\pi}^0 (0.\cos nx)\, dx + \int_0^{\pi}(\pi.\cos nx)\, dx\right] = 0$$

$$b_n = \frac{1}{\pi}\left[\int_{-\pi}^0 (0.\sin nx)\, dx + \int_0^{\pi}(\pi.\sin nx)\, dx\right] = -\frac{(-1)^n}{n} + \frac{1}{n}$$

Therefore, $b_n = 0$ if 'n' is even and $b_n = \dfrac{2}{\pi}$ if 'n' is odd.

Therefore, $f(x) = \dfrac{\pi}{2} + 2\left[\dfrac{\sin x}{1} + \dfrac{\sin 3x}{3} + \dfrac{\sin 5x}{5} + \cdots\right]$.

6)   Find the Fourier Series for f(x) in $(\pi, -\pi)$ if $f(x) = \begin{cases} x-1, -\pi < x < 0 \\ x+1,\ 0 < x < \pi \end{cases}$

Sol: Given: $f(x) = \begin{cases} x-1, -\pi < x < 0 \\ x+1,\ 0 < x < \pi \end{cases}$

We know that $f(x) = \dfrac{a_0}{2} + \sum_{n=1}^{\infty}(a_n \cos nx + b_n \sin nx)$

Now, f(x) is an odd function. So, $a_0 = a_n = 0$.

Therefore, $f(x) = \sum_{n=1}^{\infty}(b_n \sin nx)$

$$b_n = \dfrac{1}{\pi}\int_{-\pi}^{\pi} f(x)\sin nx\,dx = \dfrac{2}{\pi}\int_{0}^{\pi} f(x)\sin nx\,dx$$

$$= \dfrac{2}{\pi}\int_{0}^{\pi}(x+1)\sin nx\,dx = \left[\left[-\dfrac{\cos nx}{n}\right](x+1) - \left[-\dfrac{\sin nx}{n^2}\right](1)\right]_{0}^{\pi}$$

$$= \dfrac{2}{\pi}\left[\dfrac{-(-1)^n(\pi+1)+1}{n}\right]$$

Therefore, $b_n = -\dfrac{2}{n}$ if 'n' is even and $b_n = \dfrac{2}{\pi}\left[\dfrac{\pi+2}{n}\right]$ if 'n' is odd.

Therefore, $f(x) = \dfrac{2}{\pi}\left[\dfrac{\pi+2}{1}\right]\sin x - \dfrac{2}{2}\sin 2x + \dfrac{2}{\pi}\left[\dfrac{\pi+2}{3}\right]\sin 3x - \cdots$

7)   Evaluate $\iiint \dfrac{1}{(x+y+z+1)^3}\,dx\,dy\,dz$ over the region bounded by the plane $x = 0,\ y = 0,\ z = 0,\ x + y + z = 1$.

Sol: $I = \iiint \dfrac{1}{(x+y+z+1)^3}\,dx\,dy\,dz$

$$= \int_{0}^{1}\int_{0}^{1-x}\int_{0}^{1-x-y}(x+y+z+1)^{-3}\,dz\,dy\,dx$$

$$= \int_0^1 \int_0^{1-x} \left[\frac{(x+y+z+1)^{-2}}{-2} \right]_0^{1-x-y} dy\, dx$$

$$= \int_0^1 \int_0^{1-x} \left[-\frac{1}{8} + \frac{(x+y+1)^{-2}}{2} \right] dy\, dx$$

$$= \int_0^1 \left[-\frac{y}{8} - \frac{(x+y+1)^{-1}}{2} \right]_0^{1-x} dx$$

$$= \int_0^1 \left[-\frac{1-x}{8} - \frac{1}{4} + \frac{1}{2(1+x)} \right] dx$$

$$= \int_0^1 \left[-\frac{3}{8} + \frac{x}{4} + \frac{1}{2(1+x)} \right] dx$$

$$= \left[-\frac{3x}{8} + \frac{x^2}{16} + \frac{\log(1+x)}{2} \right]_0^1$$

$$= -\frac{5}{16} + \frac{1}{2}\log 2.$$

10 MARKS

---

1) Find the Fourier Series for f(x) if $f(x) = \begin{cases} -x, & -\pi \leq x \leq 0 \\ x, & 0 \leq x \leq \pi \end{cases}$

Sol: Given: $f(x) = \begin{cases} -x, & -\pi \leq x \leq 0 \\ x, & 0 \leq x \leq \pi \end{cases}$

Note that the function can also be expressed as $f(x) = |x|, -\pi \leq x \leq \pi$

Now, f(x) is even. Therefore, $b_n = 0$. Hence Fourier Series $f(x) = \frac{a_0}{2} + \sum_{n=1}^{\infty} a_n \cos nx$

$$a_0 = \frac{1}{\pi} \int_{-\pi}^{\pi} f(x)\, dx = \frac{2}{\pi} \int_0^{\pi} f(x)\, dx$$

$$= \frac{2}{\pi} \int_0^{\pi} x \, dx \qquad = \pi$$

$$a = \frac{1}{\pi} \int_{-}^{\pi} x \cos nx \, dx = \frac{2}{\pi} \int_0^{\pi} x \cos nx \, dx$$

$$= \frac{2}{\pi} \left[\frac{\sin nx}{n}(x) - \left[\frac{\cos nx}{n^2} \right] 1 \right]_0^{\pi}$$

$$= \frac{2}{\pi} \left[\frac{(-1)^{n-1}}{n^2} \right]$$

Therefore, $a_n = 0$ if 'n' is even and $a_n = -\frac{4}{\pi}.\frac{1}{n^2}$ if 'n' is odd.

Therefore, $f(x) = \frac{2}{\pi} - \frac{4}{\pi} \left[\frac{\cos x}{1^2} + \frac{\cos 3x}{3^2} + \frac{\cos 5x}{5^2} + \cdots \right]$.

2) Obtain a Fourier Series for the function $x + x^2$ in the interval $-\pi \leq x \leq \pi$. Hence show that $\frac{\pi^2}{12} = 1 - \frac{1}{2^2} + \frac{1}{3^2} - \cdots$

Sol: Given: $f(x) = x + x^2$

We know that $f(x) = \frac{a_0}{2} + \sum_{n=1}^{\infty} \left(a_n \cos nx + b_n \sin nx \right)$

$$a_0 = \frac{1}{\pi} \int_{-\pi}^{\pi} (x + x^2) \, dx = \frac{1}{\pi} \int_{-\pi}^{\pi} x \, dx + \frac{1}{\pi} \int_{-\pi}^{\pi} x^2 \, dx$$

$$= 0 + \frac{2}{\pi} \int_0^{\pi} x^2 \, dx \qquad = \frac{2}{\pi} \left[\frac{x^3}{3} \right]_0^{\pi} \qquad = \frac{2\pi^2}{3}$$

$$a_n = \frac{1}{\pi} \int_{-\pi}^{\pi} (x + x^2) \cos nx \, dx$$

$$= \frac{1}{\pi} \int_{-\pi}^{\pi} x \cos nx \, dx + \frac{1}{\pi} \int_{-\pi}^{\pi} x^2 \cos nx \, dx$$

$$= 0 + \frac{2}{\pi} \int_0^\pi x^2 \cos nx \, dx$$

$$= \frac{2}{\pi} \left[\frac{\sin nx}{n}(x^2) - \frac{-\cos nx}{n^2}(2x) + \frac{-\sin nx}{n^3}(2) \right]_0^\pi$$

$$= \frac{4}{\pi} \left[\frac{\cos nx}{n^2} \cdot \pi - 0 \right]$$

$$= \frac{4(-1)^n}{n^2}$$

$$b_n = \frac{1}{\pi} \int_{-\pi}^\pi (x + x^2) \sin nx \, dx$$

$$= \frac{1}{\pi} \int_{-\pi}^\pi x \sin nx \, dx + \frac{1}{\pi} \int_{-\pi}^\pi x^2 \sin nx \, dx$$

$$= \frac{2}{\pi} \int_0^\pi x \sin nx \, dx + 0$$

$$= \frac{2}{\pi} \left[\frac{-\cos nx}{n}(x) + \frac{-\sin nx}{n^2}(1) \right]_0^\pi$$

$$= \frac{2}{\pi} \left[-\frac{\cos nx}{n} \cdot \pi \right]$$

$$= \frac{2(-1)^n}{n^2}$$

Therefore, $x + x^2 = \dfrac{\pi^2}{3} + \sum_{n=1}^\infty \left[\dfrac{(-1)^n 4 \cos nx}{n^2} - \dfrac{(-1)^n 2 \sin nx}{n} \right]$

Setting x = 0, we get $0 = \dfrac{\pi^2}{3} + \sum_{n=1}^\infty \left[\dfrac{(-1)^n 4}{n^2} \right]$

$$0 = \frac{\pi^2}{3} + 4\left[-\frac{1}{1^2} + \frac{1}{2^2} + \frac{1}{3^2} + \frac{1}{4^2} - \cdots\right]$$

Therefore, $\dfrac{1}{1^2} - \dfrac{1}{2^2} + \dfrac{1}{3^2} - \dfrac{1}{4^2} + \cdots = \dfrac{\pi^2}{12}$

3) Find the Fourier Series for f(x) in $[\pi, -\pi]$ if $f(x) = \begin{cases} -\pi, & -\pi < x < 0 \\ x, & 0 < x < \pi \end{cases}$

Sol: Given: $f(x) = \begin{cases} -\pi, & -\pi < x < 0 \\ x, & 0 < x < \pi \end{cases}$

We know that $f(x) = \dfrac{a_0}{2} + \sum_{n=1}^{\infty}(a_n \cos nx + b_n \sin nx)$

$$a_0 = \frac{1}{\pi}\left[\int_{-\pi}^{0}(-\pi)\,dx + \int_{0}^{\pi} x\,dx\right] = \frac{1}{\pi}\left[\left[-\pi(x)\right]_{-\pi}^{0} + \left[\frac{x^2}{2}\right]_{0}^{\pi}\right]$$

$$= \frac{1}{\pi}\left[-\pi^2 + \frac{\pi^2}{2}\right] \qquad = \frac{-\pi}{2}$$

$$a_n = \frac{1}{\pi}\left[\int_{-\pi}^{0}(-\pi)\cos nx\,dx + \int_{0}^{\pi} x\cos nx\,dx\right]$$

$$= -\left[\frac{\sin nx}{n}\right]_{-\pi}^{0} + \frac{1}{\pi}\left[\left[\frac{\sin nx}{n}\right]x - \left[\frac{-\cos nx}{n^2}\right]1\right]_{0}^{\pi}$$

$$= 0 + \frac{1}{\pi}\left[0 + \frac{\cos n\pi}{n^2} - \frac{1}{n^2}\right]$$

$$= \frac{1}{\pi}\left[\frac{(-1)^n - 1}{n^2}\right]$$

Therefore, $a_n = 0$ if 'n' is even and $a_n = -\dfrac{2}{\pi}\cdot\dfrac{1}{n^2}$ if 'n' is odd.

$$b_n = \frac{1}{\pi}\left[\int_{-\pi}^{0}(-\pi)\sin nx\,dx + \int_{0}^{\pi} x\sin nx\,dx\right]$$

$$= -\left[-\frac{\cos nx}{n}\right]_{-\pi}^{0} + \frac{1}{\pi}\left[\left[-\frac{\cos nx}{n}\right]x - \left[\frac{-\sin nx}{n^2}\right]1\right]_{0}^{\pi}$$

$$= \frac{1}{n} - \frac{\cos n\pi}{n} + \frac{1}{\pi}\left[-\frac{\pi\cos n\pi}{n} + 0\right]$$

$$= \frac{1 - 2(-1)^n}{n}$$

Therefore, $b_n = -\dfrac{1}{n}$ if 'n' is even and $b_n = \dfrac{3}{n}$ if 'n' is odd.

Therefore,

$$f(x) = -\frac{\pi}{4} - \frac{2}{\pi}\left[\frac{\cos x}{1^2} + \frac{\cos 3x}{3^2} + \frac{\cos 5x}{5^2} + \cdots\right] + \left[\frac{3\sin x}{1} - \frac{\sin 2x}{2} + \frac{3\sin 3x}{3} - \cdots\right]$$

TRIGONOMETRY

2 MARKS

1) Write the formula for $\sin n\theta$

Sol: $\sin n\theta = nc_1 \cos^{n-1}\theta \sin\theta - nc_3 \cos^{n-3}\theta \sin^3\theta + nc_5 \cos^{n-5}\theta \sin^5\theta \ldots$

2) Write the formula for $\cos n\theta$.

Sol: $\cos n\theta = \cos^n\theta - nc_2 \cos^{n-2}\theta \sin^2\theta + nc_4 \cos^{n-4}\theta \sin^4\theta \ldots$

3) Write the formula for $\tan n\theta$.

Sol: $\tan n\theta = {}^nC1 \tan\theta - {}^nC_3 \tan^3\theta + {}^nC_5 \tan^5\theta - \ldots / 1 - {}^nC_2 \tan^2\theta + {}^nC_4 \tan^4\theta - \ldots$

4) Expand $\tan 7\theta$ in terms of θ.

Sol: W.K.T $\tan n\theta = {}^nC_1 \tan\theta - {}^nC_3 \tan^3\theta + {}^nC_5 \tan^5\theta - \ldots / 1 - {}^nC_2 \tan^2\theta + {}^nC_4 \tan^4\theta - \ldots$

$\tan 7\theta = {}^7C_1 \tan\theta - {}^7C_3 \tan^3\theta + {}^7C_5 \tan^5\theta - {}^7C_7 \tan^7\theta / 1 - {}^7C_2 \tan^2\theta + {}^7C_4 \tan^4\theta - {}^7C_6 \tan^6\theta$

$\tan 7\theta = 7\tan\theta - 35\tan^3\theta + 21\tan^5\theta - \tan^7\theta / 1 - 21\tan^2\theta + 35\tan^4\theta - 7\tan^6\theta$.

5) Expansion of $\cos^n\theta$ and $\sin n\theta$ interms of sines and cosines of multiples of θ.

Sol: Let $x = \cos\theta + i\sin\theta \implies \dfrac{1}{x} = \cos\theta - i\sin\theta$

Subtracting, $x - \dfrac{1}{x} = 2i \sin \theta$

Adding, $x + \dfrac{1}{x} = 2 \cos \theta$

$x^n = (\cos \theta + i \sin \theta)n$ ➜ $\dfrac{1}{x^n} = (\cos \theta - i \sin \theta)^n$

Subtracting, $x^n - \dfrac{1}{x^n} = 2i \sin n\theta$

Adding, $x^n + \dfrac{1}{x^n} = 2 \cos n\theta$

5 MARKS

1) If $\dfrac{\tan\theta}{\theta} = \dfrac{2524}{2523}$, then find θ approximately.

Sol: Given: $\dfrac{\tan\theta}{\theta} = \dfrac{2524}{2523}$

i.e., $\dfrac{\theta + \dfrac{\theta^3}{3} + \dfrac{2\theta^5}{15} + \cdots}{\theta} = \dfrac{2524}{2523}$

Therefore, $\theta + \dfrac{\theta^2}{3} + \cdots = \dfrac{2524}{2523}$

$\dfrac{\theta^2}{3} = \dfrac{1}{2523}$ nearly

$\theta^2 = \dfrac{1}{841}$ nearly

$\theta = \dfrac{1}{29}$ nearly

$\theta = 1°58'$ approximately.

2) If $u = \log \tan \left[\dfrac{\pi}{4} + \dfrac{\alpha}{2}\right]$, then prove that $\cosh u = \sec \alpha$.

Sol: Given : $u = \log \tan \left[\dfrac{\pi}{4} + \dfrac{\alpha}{2}\right]$

$u = \log \tan \left[\left[\dfrac{\pi}{4} + \dfrac{\alpha}{2}\right]\right]$

$$= \log\left[\frac{1+\tan\left[\dfrac{\alpha}{2}\right]}{1-\tan\left[\dfrac{\alpha}{2}\right]}\right]$$

Taking exponential on both sides,

$$e^{u} = \frac{1+\tan\left[\dfrac{\alpha}{2}\right]}{1-\tan\left[\dfrac{\alpha}{2}\right]} \quad \blacktriangleright (1)$$

Similarly, $\quad e^{-u} = \dfrac{1-\tan\left[\dfrac{\alpha}{2}\right]}{1+\tan\left[\dfrac{\alpha}{2}\right]} \quad \blacktriangleright (2)$

We know that cosh $u = \dfrac{e^{u}+e^{-u}}{2}$

Substituting (1) and (2) in the above equation, we get cosh

$$u = \frac{\dfrac{1+\tan\left[\dfrac{\alpha}{2}\right]}{1-\tan\left[\dfrac{\alpha}{2}\right]} + \dfrac{1-\tan\left[\dfrac{\alpha}{2}\right]}{1+\tan\left[\dfrac{\alpha}{2}\right]}}{2}$$

$$2\cosh u = \frac{1+\tan\left[\dfrac{\alpha}{2}\right]}{1-\tan\left[\dfrac{\alpha}{2}\right]} + \frac{1-\tan\left[\dfrac{\alpha}{2}\right]}{1+\tan\left[\dfrac{\alpha}{2}\right]}$$

$$= \frac{\left(1+\tan\left[\dfrac{\alpha}{2}\right]\right)^{2} + \left(1-\tan\left[\dfrac{\alpha}{2}\right]\right)^{2}}{\left(1-\tan\left[\dfrac{\alpha}{2}\right]\right)\left(1+\tan\dfrac{\alpha}{2}\right)}$$

$$= \frac{2\left(1 + \tan^2\left[\dfrac{\alpha}{2}\right]\right)}{1 - \tan^2\left[\dfrac{\alpha}{2}\right]}$$

$$\cosh u = \frac{\dfrac{\sin^2 \dfrac{\alpha}{2} + \cos^2 \dfrac{\alpha}{2}}{\cos^2 \dfrac{\alpha}{2}}}{\dfrac{\cos^2 \dfrac{\alpha}{2} - \sin^2 \dfrac{\alpha}{2}}{\cos^2 \dfrac{\alpha}{2}}}$$

$$\cosh u = \frac{1}{2\cos^2 \dfrac{\alpha}{2} - 1}$$

$$\cosh u = \frac{1}{\cos \alpha}$$

$$\cosh u = \sec \alpha.$$

3) Prove that $- 64 \sin^7 \theta = \sin 7\theta - 7 \sin 5\theta + 21 \sin 3\theta - 35 \sin \theta.$

Sol: Let $x = \cos \theta + i \sin \theta \;\rightarrow\; \dfrac{1}{x} = \cos \theta - i \sin \theta$

Subtracting, $x - \dfrac{1}{x} = 2i \sin \theta$

Adding, $x - \dfrac{1}{x} = 2 \cos \theta$

$$x^n = (\cos \theta + i \sin \theta)^n \;\rightarrow\; \dfrac{1}{x^n} = (\cos \theta - i \sin \theta)^n$$

Subtracting, $x^n - \dfrac{1}{x^n} = 2i \sin n\theta$

Adding, $x^n + \dfrac{1}{x^n} = 2\cos n\theta$

Now, consider $\left[x - \dfrac{1}{x}\right]^7$

$$\left[x - \frac{1}{x}\right]^7 = {}^7C_0 x^7 - {}^7C_1 x^5 + {}^7C_2 x^3 - {}^7C_3 x + {}^7C_4 \frac{1}{x} - {}^7C_5 \frac{1}{x^3} + {}^7C_6 \frac{1}{x^5} - {}^7C_7 \frac{1}{x^7}$$

Therefore,

$(2i \sin \theta)^7 = 2i \sin 7\theta - 7(2i \sin 5\theta) + 21(2i \sin 3\theta) - 35(2i \sin\theta)$

$2^7 i^7 \sin^7 \theta = 2i(\sin 7\theta - 7 \sin 5\theta + 21 \sin 3\theta - 35 \sin \theta)$

Cancelling 2i,

$-64 \sin^7 \theta = \sin 7\theta - 7 \sin 5\theta + 21 \sin 3\theta - 35 \sin \theta.$

4) If $\tan\left[\dfrac{x}{2}\right] = \tanh\left[\dfrac{y}{2}\right]$, then prove that (i) $\sinh y = \tan x$.

   (ii) $2y = \log \tan\left[\dfrac{\pi}{4} + \dfrac{x}{2}\right]$.

Sol: Given: $\tan\left[\dfrac{x}{2}\right] = \tanh\left[\dfrac{y}{2}\right]$ ➡   (1)

$$\tan x = \frac{2\tan\left[\dfrac{x}{2}\right]}{1 - tan^2\left[\dfrac{x}{2}\right]}$$

$$= \frac{2\tanh\left[\dfrac{y}{2}\right]}{1 - tanh^2\left[\dfrac{y}{2}\right]}$$

$$= \frac{2\sinh\left[\dfrac{y}{2}\right]}{\cosh\left[\dfrac{y}{2}\right]} . cosh^2\left[\dfrac{y}{2}\right]$$

$$= 2\,\sinh\left[\frac{y}{2}\right]\cosh\left[\frac{y}{2}\right]$$

$$= \sinh y$$

Also from (1), $\dfrac{y}{2} = \tanh^{-1}\left[\tan\dfrac{x}{2}\right]$

$$\frac{y}{2} = \frac{1}{2}\log\left[\frac{1+\tan\dfrac{x}{2}}{1-\tan\dfrac{x}{2}}\right]$$

$$y = \log\tan\left[\frac{\pi}{4}+\frac{x}{2}\right].$$

5) Show that $\cos8\theta = 128\cos8\theta - 256\cos6\theta + 160\cos4\theta - 32\cos^2\theta + 1$

Sol: $\cos n\theta = \cos^n\theta - nc_2\cos^{n-2}\theta\sin^2\theta + nc_4\cos^{n-4}\theta\sin^4\theta\ldots\ldots$

$\cos8\theta = \cos^8\theta - 8c_2\cos^6\theta\sin^2\theta + 8c_4\cos^4\theta\sin^4\theta - 8c_6\cos^2\theta\sin^6\theta + 8c_8\sin^8\theta$

$\cos8\theta = \cos^8\theta - 28\cos^6\theta\,(1-\cos^2\theta) + 70\cos^4\theta\,(1-\cos^2\theta)^2 - 28\cos^2\theta\,(1-\cos^2\theta)^3 + 8c_8\,(1-\cos^2\theta)^4$

$\cos8\theta = \cos^8\theta - 28\cos^6\theta\,(1-\cos^2\theta) + 70\cos^4\theta\,(1-2\cos^2\theta+\cos^4\theta) - 28\cos^2\theta\,(1-3\cos^2\theta+3\cos^4\theta-\cos^6\theta) + \cos^8\theta - 4\cos^6\theta + 6\cos^4\theta - 4\cos^2\theta + 1.$

$\cos8\theta = 128\cos^8\theta - 256\cos^6\theta + 160\cos^4\theta - 32\cos^2\theta + 1.$

6) Express $\cos6\theta$ in terms of $\cos\theta$

Sol: $\cos n\theta = \cos^n\theta - nc_2\cos^{n-2}\theta\sin^2\theta + nc_4\cos^{n-4}\theta\sin^4\theta\ldots\ldots$

$\cos6\theta = \cos^6\theta - 6c_2\cos^4\theta\sin^2\theta + 6c_4\cos^2\theta\sin^4\theta - 6c_6\sin^6\theta$

$\cos6\theta = \cos^6\theta - 15\cos^4\theta\,(1-\cos^2\theta) + 15\cos^2\theta\,(1-\cos^2\theta)^2 - (1-\cos^2\theta)^3$

$\cos 6\theta = \cos^6\theta - 15\cos^4\theta\,(1-\cos^2\theta) + 15\cos^2\theta\,(1 - 2\cos^2\theta + \cos^4\theta) - (1 - 3\cos^2\theta + 3\cos^4\theta - \cos^6\theta)$

$\cos 6\theta = 32\cos^6\theta - 48\cos^4\theta + 18\cos^2\theta - 1.$

7) Expand $\sin 6\theta$ in powers of $\cos\theta$ and $\sin\theta$.

Sol: $\sin n\theta = nc_1\cos^{n-1}\theta\,\sin\theta - nc_3\cos^{n-3}\theta\,\sin^3\theta\ldots\ldots$

$\sin 6\theta = 6c_1\cos^5\theta\,\sin\theta - 6c_3\cos^3\theta\,\sin^3\theta + 6c_5\cos\theta\,\sin^5\theta$

$\sin 6\theta = 6\cos^5\theta\,\sin\theta - 20\cos^3\theta\,\sin^3\theta + 6\cos\theta\,\sin^5\theta.$

8) Express $\cos 6\theta$ in series of $\cos\theta$ and $\sin\theta$

Sol: $\cos n\theta = \cos^n\theta - nc_2\cos^{n-2}\theta\,\sin^2\theta + nc_4\cos^{n-4}\theta\,\sin^4\theta\ldots\ldots$

$\cos 6\theta = \cos^6\theta - 6c_2\cos^4\theta\,\sin^2\theta + 6c_4\cos^2\theta\,\sin^4\theta - 6c_6\sin^6\theta$

$\cos 6\theta = \cos^6\theta - 15\cos^4\theta\,\sin^2\theta + 15\cos^2\theta\,\sin^4\theta - \sin^6\theta$

9) Expand $\sin 5\theta$ as a polynomial in $\sin\theta$.

Sol: $\sin n\theta = nc_1\cos^{n-1}\theta\,\sin\theta - nc_3\cos^{n-3}\theta\,\sin^3\theta\ldots\ldots$

$\sin 5\theta = 5c_1\cos^4\theta\,\sin\theta - 5c_3\cos^2\theta\,\sin^3\theta + 5c_5\sin^5\theta$

$\sin 5\theta = 5(1 - \sin^2\theta)^2\,\sin\theta - 10(1 - \sin^2\theta)\,\sin^3\theta + \sin^5\theta$

$\sin 5\theta = 5(1 - 2\sin^2\theta + \sin^4\theta)\,\sin\theta - 10\sin^3\theta + 10\sin^5\theta + \sin^5\theta$

$\sin 5\theta = (5 - 10\sin^2\theta + 5\sin^4\theta)\,\sin\theta - 10\sin^3\theta + 10\sin^5\theta + \sin^5\theta$

$\sin 5\theta = 5\sin\theta - 10\sin^3\theta + 5\sin^5\theta - 10\sin^3\theta + 10\sin^5\theta + \sin^5\theta$

$\sin 5\theta = 16\sin^5\theta - 20\sin^3\theta + 5\sin\theta.$

10) Prove that $\dfrac{1+\tanh x}{1-\tanh x} = \cosh 2x + \sinh 2x.$

Sol: $\cos 2x = \dfrac{1-\tan^2 x}{1+\tan^2 x}$

Change x into ix, $\cos(2ix) = \dfrac{1-(\tan ix)^2}{1+(\tan ix)^2}$

$$i\cos 2x = \frac{1-i^2\tanh^2 x}{1+i^2\tanh^2 x}$$

$$\cosh 2x = \frac{1+\tanh^2 x}{1-\tanh^2 x} \quad\longrightarrow\quad (1)$$

$$\sin 2x = \frac{2\tan x}{1+\tan^2 x}$$

Change x into ix, $\sin(2ix) = \dfrac{2\tan ix}{1+(\tan ix)^2}$

$$i\sinh 2x = \frac{2i\tanh x}{1+i2\tanh^2 x}$$

$$\sinh 2x = \frac{2\tanh x}{1-\tanh^2 x}$$

Adding (1) and (2), we get

$$\cosh 2x + \sinh 2x = \frac{1+\tanh^2 x}{1-\tanh^2 x} + \frac{2\tanh x}{1-\tanh^2 x}$$

$$= \frac{(1+\tanh x)^2}{1-\tanh^2 x}$$

$$= \frac{1+\tanh x}{(1-\tanh x}.$$

11) If $\dfrac{\sin\theta}{\theta} = \dfrac{5045}{5046}$, then find θ approximately.

Sol: Given $\dfrac{\sin\theta}{\theta} = \dfrac{5045}{5046}$

i.e., $\dfrac{1-\dfrac{\theta^2}{3!}+\dfrac{\theta^4}{5!}}{\theta} = \dfrac{5045}{5046}$

Therefore, $1-\dfrac{\theta^2}{6} = 1-\dfrac{1}{5046}$

$$\frac{\theta^2}{6} = \frac{1}{5046} \text{ nearly}$$

$$\frac{\theta^2}{6} = \frac{1}{5046} \text{ nearly}$$

$$\theta = \frac{1}{29} \text{ nearly}$$

$\theta = 1°58'$ approximately.

10 MARKS

1) If $\log[\sin(\theta + i\phi)] = A + iB$, then prove that $2e^{2A} = \cosh 2\phi - \cos 2\theta$.

Sol: Given: $\log[\sin(\theta + i\phi)] = A + iB$

$$\log[\sin\theta \cosh\phi + i\cos\theta \sinh\phi] = A + iB$$

$$\frac{1}{2}\log[\sin^2\theta \cosh^2\phi + \cos^2\theta \sinh^2\phi] + i\tan^{-1}\left[\frac{\cos\theta \sinh\phi}{\sin\theta \cosh\phi}\right] = A + iB \quad \rightarrow (1)$$

Equating real parts, $\log[\sin^2\theta \cosh^2\phi + i\cos^2\theta \sinh^2\phi] = 2A$

$$\frac{1 - \cos 2\theta}{2}[\cosh^2\phi] + \frac{1 + \cos 2\theta}{2}[\sinh^2\phi] = e^{2A}$$

$(\cosh^2\phi + \sinh^2\phi) - \cos 2\theta (\cosh^2\phi - \sinh^2\phi) = 2e^{2A} \cosh 2\phi - \cos 2\theta = 2e^{2A}$

Hence proved.

2) Expand $\dfrac{\sin 7\theta}{\sin\theta}$ as a polynomial in $\cos\theta$.

Sol: By Demoivre's theorem, $(\cos\theta + i\sin\theta)^7 = \cos 7\theta + i\sin 7\theta \quad \rightarrow (1)$

By Binomial theorem,

$$(\cos\theta + i\sin\theta)^7 = \cos^7\theta + {}^7C_1 \cos^6\theta\, i\sin\theta + {}^7C_2\cos^5\theta\, i^2 \sin^2\theta +$$
$${}^7C_3 \cos^4\theta\, i^3 \sin^3 + \theta \qquad {}^7C_4 \cos^3\theta\, i^4 \sin^4\theta + {}^7C_5$$

$$cos^2\ \theta\ i^5\ sin^5\ \theta + {}^7C_6\ cos\ \theta\ i^6\ sin^6\ \theta + {}^7C_7\ i^7\ sin^7\theta$$

$$= cos^7\ \theta - 21\ cos^5\ \theta\ sin^2\ \theta + 35\ cos^3\ \theta\ sin^4\ \theta - 7\ cos\theta\ sin^6\ \theta + i$$

$$[7cos^6\ \theta\ sin\ \theta - 35\ cos^4\ \theta\ sin^3\ \theta + 21\ cos^2\ \theta\ sin^5\ \theta - sin^7\ \theta] \quad\rightarrow\quad (2)$$

Equating imaginary parts in (1) and (2),

$$sin\ 7\theta = 7\ cos^6\ \theta\ sin\ \theta - 35\ cos^4\ \theta\ sin^3\ \theta + 21\ cos^2\ \theta\ sin^5\ \theta - sin^7\ \theta$$

Dividing by sin , we get

$$\frac{sin\ 7\theta}{sin\ \theta} = 7cos^6\theta - 35\ cos^4\theta\ sin^2\theta + 21\ cos^2\theta\ sin^4\theta - sin^6\theta$$

$$= 7\ cos^6\ \theta - 35\ cos^4\ \theta\ (1 - cos^2\ \theta) + 21\ cos^2\ \theta\ (1 - cos^2\ \theta)^2 - (1 - cos^2\ \theta)^3$$

$$= 7\ cos^6\ \theta - 35\ cos^4\ \theta\ (1 - cos^2\ \theta) + 21\ cos^2\ \theta\ (1 - 2cos^2\ \theta + cos^4\ \theta) - (1 - 3cos^2\ \theta + 3\ cos^4\ \theta - cos^6\ \theta)$$

$$= 64\ cos^6\ \theta - 80\ cos^4\ \theta + 24\ cos^2\ \theta - 1.$$

3) Prove that $- 32\ sin^6\theta = cos6\theta - 6cos4\theta + 15\ cos\ 2\theta - 10$.

Sol: Let $x = cos\ \theta + i\ sin\ \theta \rightarrow \dfrac{1}{x} = cos\ \theta - i\ sin\ \theta$

Subtracting, $x - \dfrac{1}{x} = 2i\ sin\ \theta$

Adding, $x + \dfrac{1}{x} = 2\ cos\ \theta$

$x^n = (cos\ \theta + i\ sin\ \theta)^n \rightarrow \dfrac{1}{x^n} = (cos\ \theta - i\ sin\ \theta)^n$

Subtracting, $x^n - \dfrac{1}{x^n} = 2i\ sin\ n\ \theta$

Adding, $x^n + \dfrac{1}{x^n} = 2\ cos\ n\ \theta$

Now, consider $\left[x - \dfrac{1}{x}\right]^6$

$$\left[x - \frac{1}{x}\right]^6 = x^6 - {}^6C_1\ x^5\ \frac{1}{x} + {}^6C_2\ x^4\ \frac{1}{x^2} - {}^6C_3\ x^3\ \frac{1}{x^3} + {}^6C_4\ x^2\ \frac{1}{x^4} - {}^6C_5\ x\ \frac{1}{x^5} + {}^6C_6\ \frac{1}{x^6}$$

Therefore, $(2i \sin \theta)^6 = 2 \cos 6\theta - 6(2 \cos 4\theta) + 15(2 \cos 2\theta) - 20$

$2^6 \, i^6 \sin^6 \theta = 2 \cos 6\theta - 12 \cos 4\theta + 30 \cos 2\theta - 20$

Cancelling 2 on both sides,

$-2^6 \sin^6 \theta = \cos 6\theta - 6 \cos 4\theta + 15 \cos 2\theta - 10$

$\mathrm{Sin}^6 \theta = -\dfrac{1}{32} (\cos 6\theta - 6 \cos 4\theta + 15 \cos 2\theta - 10).$

4) Prove that $\cos^5\theta \sin^4\theta = \dfrac{1}{2^8} (\cos 9\theta + \cos 7\theta - 4 \cos 5\theta - 4\cos 3\theta + 6\cos\theta).$

Sol: Let $x = \cos \theta + i \sin \theta$ $\Rightarrow$ $\dfrac{1}{x} = \cos \theta - i \sin \theta$

Subtracting, $x - \dfrac{1}{x} = 2i \sin \theta$

Adding, $x + \dfrac{1}{x} = 2 \cos \theta$

$x^n = (\cos \theta + i \sin \theta)^n$ $\Rightarrow$ $\dfrac{1}{x^n} = (\cos \theta - i \sin \theta)^n$

Subtracting, $x^n - \dfrac{1}{x^n} = 2i \sin n \theta$

Adding, $x^n + \dfrac{1}{x^n} = 2 \cos n \theta$

Consider $(2 \cos\theta)(2i\sin\theta) = \left(x + \dfrac{1}{x}\right)\left(x - \dfrac{1}{x}\right)$

$$(2 \cos\theta)^5 (2i\sin\theta)^4 \left(x + \dfrac{1}{x}\right)^5 \left(x - \dfrac{1}{x}\right)^4$$

$$(2 \cos\theta)^5 (2i\sin\theta)^4 \left(x + \dfrac{1}{x}\right) \left(x^2 - \dfrac{1}{x^2}\right)^4$$

$$(2 \cos\theta)^5 (2i\sin\theta)^4 \left(x + \dfrac{1}{x}\right) \left(x^8 - 4x^4 + 6 - 4\dfrac{1}{x^4} + \dfrac{1}{x^8}\right)$$

$2^5 \cos^5\theta \; 2^4 i^4 \sin^4\theta = 2\cos 9\theta + 2\cos 7\theta - 4.\,2\cos 5\theta - 4.\,2\cos 3\theta + 6.\,2\cos\theta$

$2^8 \cos^5\theta \sin^4\theta = \cos 9\theta + \cos 7\theta - 4\cos 5\theta - 4\cos 3\theta + 6\cos\theta$

$\cos^5\theta \sin^4\theta = \dfrac{1}{2^8}(\cos 9\theta + \cos 7\theta - 4\cos 5\theta - 4\cos 3\theta + 6\cos\theta).$

LAPLACE TRANSFORMS

2 MARKS

1) Find $L(\cos^3 t)$.

Sol: $\cos 3A = 4\cos^3 A - 3\cos A$

$$\cos^3 A = \frac{1}{4}[\cos 3A + 3\cos A]$$

Therefore, $L(\cos^3 t) = \frac{1}{4} L[\cos 3t + 3\cos t]$

$$= \frac{1}{4}[L(\cos 3t) + L(3\cos t)]$$

$$= \frac{1}{4}\left[\frac{s}{s^2 + 9} + 3\left[\frac{s}{s^2 + 1}\right]\right].$$

2) Find $L(e^t + \sin t)$.

Sol: $L(e^t + \sin t) = L(e^t) + L(\sin t) = \frac{1}{s-1} + \frac{1}{s^2 + 1^2}.$

3) Find $L[\sin 4t \sin 6t]$.

Sol: We know that $2\sin A \sin B = [\cos(A - B) - \cos(A + B)]$

$$\sin A \sin B = -\frac{1}{2}[\cos(A + B) - \cos(A - B)]$$

$$L(\sin 4t \sin 6t) = -\frac{1}{2}L(\cos(4t + 6t) - \cos(4t - 2t))$$

$$= -\frac{1}{2} L(\cos 10t - \cos(-2t))$$

$$= -\frac{1}{2}\left[\frac{s}{s^2+10^2} - \frac{s}{s^2+(-2)^2}\right]$$

$$= -\frac{1}{2}\left[\frac{s}{s^2+100} - \frac{2}{s^2+4}\right].$$

4) Find $L^{-1}\left[\dfrac{1}{s-3} + \dfrac{1}{s} + \dfrac{s}{s^2-4}\right].$

Sol:

$$L^{-1}\left[\frac{1}{s-3}+\frac{1}{s}+\frac{s}{s^2-4}\right] = L^{-1}\left[\frac{1}{s-3}\right] + L^{-1}\left[\frac{1}{s}\right] + L^{-1}\left[\frac{s}{s^2-4}\right] = e^{3t} + 1 + \cosh 2t.$$

5) Find $L^{-1}\left[\dfrac{s^2}{(s-4)^4}\right].$

Sol: $\quad L^{-1}\left[\dfrac{s^2}{(s-4)^4}\right] = L^{-1}\left[\dfrac{[(s-4)+4]^2}{(s-4)^4}\right]$

$$= e^{4t}\, L^{-1}\left[\frac{(s+4)^2}{s^2}\right] \qquad \text{(by replacing } (s-4) \text{ by } s)$$

$$= e^{4t}\, L^{-1}\left[\frac{s^2+8s+16}{s^4}\right]$$

$$= e^{4t}\, L^{-1}\left[\frac{1}{s^2}+\frac{8}{s^3}+\frac{16}{s^4}\right] = e^{4t}\, L^{-1}\left[t+8\left[\frac{t^2}{2!}\right]+16\left[\frac{t^3}{3!}\right]\right].$$

6) Define Inverse Laplace Transform.

Sol: If $L[f(t)] = F(s)$, then $F(t)$ is called the inverse laplace transform of $F(s)$ and is denoted by $L^{-1}[F(s)]$.

7) Find L(t sinh at).

Sol: We know that $L(\sinh at) = \dfrac{a}{s^2 - a^2}$

$$L(t \sinh at) = -\frac{d}{ds}\left[\frac{a}{s^2 - a^2}\right] = \frac{2as}{[s^2 - a^2]^2}.$$

8) Find $L^{-1}\left[\dfrac{s}{s^2 + 2s + 10}\right]$.

Sol: $L^{-1}\left[\dfrac{s}{s^2 + 2s + 10}\right] = L^{-1}\left[\dfrac{s}{(s+1)^2 + 3^2}\right]$

$$= L^{-1}\left[\frac{(s+1)-1}{(s+1)^2 + 3^2}\right]$$

$$= e^{-t} L^{-1}\left[\frac{s}{s^2 + 3^2} - \frac{1}{s^2 + 3^2}\right]$$

$$= e^{-t}(\cos 3t - \frac{1}{3}\sin 3t).$$

5 MARKS

1) Find $L\left(\dfrac{e^{-at} - e^{-bt}}{t}\right)$.

Sol: We know that $L(e^{-at} - e^{-bt}) = \dfrac{1}{s+a} - \dfrac{1}{s+b}$

Now $L\left(\dfrac{e^{-at} - e^{-bt}}{t}\right) = \displaystyle\int_s^\infty \left[\frac{1}{s+a} - \frac{1}{s+b}\right] ds$

$$= [\log(s+a) - \log(s+b)]_s^\infty$$

$$= \left[\log\left[\frac{s+a}{s+b}\right]\right]_s^\infty$$

$$= \left[\log \left[\frac{1 + \dfrac{a}{s}}{1 + \dfrac{b}{s}} \right] \right]_{s}^{\infty}$$

$$= \log 1 - \log \left[\frac{1 + \dfrac{a}{s}}{1 + \dfrac{b}{s}} \right]$$

$$= 0 - \log \frac{s+a}{s+b} \qquad = \log \frac{s+b}{s+a}.$$

2) Find $L^{-1}\left(\dfrac{s-2}{s^2 + 2s + 2} \right)$.

Sol: $L^{-1}\left(\dfrac{s-2}{s^2 + 2s + 2} \right) = L^{-1}\left[\dfrac{(s+1)-3}{(s+1)^2 + 1^2} \right]$

$$= L^{-1}\left[\frac{(s+1)}{(s+1)^2 + 1^2} - \frac{3}{(s+1)^2 + 1^2} \right]$$

$$= e^{-t} L^{-1}\left[\frac{s}{s^2 + 1^2} - \frac{3}{s^2 + 1^2} \right]$$

$$= e^{-2t}(\cos t - 3 \sin t).$$

3) Find $L^{-1}\left(\dfrac{1}{(s-2)(s+1)^2} \right)$.

Sol: Let $\dfrac{1}{(s-2)(s+1)^2} = \dfrac{A}{s-2} + \dfrac{B}{s+1} + \dfrac{C}{(s+1)^2}$

$1 = A(s + 1)^2 + B(s - 2)(s + 1) + C(s - 2)$

Put $s = 2$, $A = \dfrac{1}{9}$

Put $s = -1$, $C = -\dfrac{1}{3}$

Comparing the coefficients of s^2, $0 = A + B$ $\rightarrow$ $B = -\dfrac{1}{9}$

Therefore, $L^{-1}\left(\dfrac{1}{(s-2)(s+1)^2} \right) = \dfrac{\frac{1}{9}}{s-2} - \dfrac{\frac{1}{9}}{s+1} - \dfrac{\frac{1}{3}}{(s+1)^2}$

$$= \frac{1}{9}e^{2t} - \frac{1}{9}e^{-t} - \frac{1}{3}e^{-t} L^{-1}\left[\frac{1}{s^2} \right]$$

$$= \frac{1}{9}e^{2t} - \frac{1}{9}e^{-t} - \frac{1}{3}e^{-t}.t.$$

4) Find $L(t^2 e^{3t} \sinh t)$.

Sol: $L(t^2 e^3t \sinh t) = L\left[t^2 e^{3t}\left[\frac{e^t - e^{-t}}{2}\right]\right]$

$$= \frac{1}{2} L(t^2 e^{4t} - t^2 e^{2t})$$

$$= \frac{1}{2}\left[\left[\frac{2!}{s^3}\right]_{s \to s-4} - \left[\frac{2!}{s^3}\right]_{s \to s-2}\right] \text{ (by LT 1)}$$

$$= \frac{1}{(s-4)^3} - \frac{1}{(s-2)^3}.$$

5) Find $L(e^{-3t} \sin t \cos t)$.

Sol: We know that $\sin t \cos t = \frac{1}{2} \sin 2t$ and $L(\sin 2t) = \frac{2}{s^2 + 2^2}$

Therefore, $L(e^{-3t} \sin t \cos t) = - L(e^{-3t} \sin 2t)$

$$= \frac{1}{2} L\left[\frac{2}{s^2 + 2^2}\right]_{s \to s+3} \text{ (by LT 1)}$$

$$= \frac{1}{2}\left[\frac{2}{(s+3)^2 + 2^2}\right].$$

6) Find $L^{-1}\left(\frac{s-3}{s^2 + 4s + 13}\right)$.

Sol: $L^{-1}\left(\frac{s-3}{s^2 + 4s + 13}\right) = = L^{-1}\left[\frac{(s+2)-5}{(s+2)^2 + 3^2}\right]$

$$= L^{-1}\left[\frac{(s+2)}{(s+2)^2 + 3^2} - \frac{5}{(s+2)^2 + 3^2}\right]$$

$$= e^{-2t} L^{-1}\left[\frac{s}{s^2 + 3^2} - \frac{5}{s^2 + 3^2}\right]$$

$$= e^{-2t}\left(\cos 3t - \frac{5}{3}\sin 3t\right).$$

7) Find $L\left(\dfrac{1-\cos at}{t}\right)$.

Sol: We know that $L(\cos at - \cos bt) = \dfrac{s}{s^2+a^2} - \dfrac{s}{s^2+b^2}$

$$L\left[\frac{\cos at - \cos bt}{t}\right] = \int_s^\infty \left[\frac{s}{s^2+a^2} - \frac{s}{s^2+b^2}\right] ds$$

$$= \frac{1}{2}[\log(s^2+a^2) - \log(s^2+b^2)]_s^\infty$$

$$= \frac{1}{2}\left[\log\frac{s^2+a^2}{s^2+b^2}\right]_s^\infty$$

$$= \frac{1}{2}\left[\log\frac{1+\dfrac{a^2}{s^2}}{1+\dfrac{b^2}{s^2}}\right]_s^\infty$$

$$= \frac{1}{2}\left[\log 1 - \frac{\log 1 + \left[\dfrac{a^2}{s^2}\right]}{\log 1 + \left[\dfrac{b^2}{s^2}\right]}\right]$$

$$= \frac{1}{2}\left[0 - \log\frac{s^2+a^2}{s^2+b^2}\right]$$

$$= \frac{1}{2}\left[\log\frac{s^2+b^2}{s^2+a^2}\right]$$

Put a = 0 and b = a in the above equation, we get

$$L\left(\frac{1-\cos at}{t}\right) = \frac{1}{2}\left[\log\frac{s^2+a^2}{s^2}\right].$$

8) Find $L^{-1}\left(\dfrac{1}{s(s+1)(s+9)}\right)$.

Sol: $L^{-1}\left(\dfrac{1}{s(s+1)(s+9)}\right) = \dfrac{A}{s} + \dfrac{B}{s+1} + \dfrac{C}{s+9}$

$$A = \left(\dfrac{1}{(s+1)(s+9)}\right)_{s=0} = \dfrac{1}{9}$$

$$B = \left(\dfrac{1}{s(s+9)}\right)_{s=-1} = -\dfrac{1}{8}$$

$$C = \left(\dfrac{1}{s(s+1)}\right)_{s=-9} = \dfrac{1}{72}$$

Therefore, $L^{-1}\left(\dfrac{1}{s(s+1)(s+9)}\right) = \dfrac{1}{9}\cdot\dfrac{1}{s} - \dfrac{1}{8}\cdot\dfrac{1}{s+1} + \dfrac{1}{72}\cdot\dfrac{1}{s+9}$

$$= \dfrac{1}{9} - \dfrac{1}{8}e^{-t} + \dfrac{1}{72}e^{-9t}.$$

10 MARKS

---

1) Solve by Laplace Transform method $\dfrac{d^2y}{dx^2} + 4\dfrac{dy}{dx} + 13y = 2e^{-x}$, given that $y(0) = 0$, $y'(0) = -1$.

Sol: Given: $\dfrac{d^2y}{dx^2} + 4\dfrac{dy}{dx} + 13y = 2e^{-x}$

Taking Laplace transforms on both sides,

$L(y'' + 4y' + 13y) = L(2e^{-x})$

i.e., $L(y'') + 4L(y') + 13L(y) = 2L(e^{-x})$

i.e., $[s^2 L(y) - sy(0) - y'(0)] + 4[s L(y) - y(0)] + 13 L(y) = \dfrac{2}{s+1}$

But, it is given that $y(0) = 0$, $y'(0) = -1$.

Therefore $[s^2 L(y) + 1] + 4s L(y) + 13 L(y) = \dfrac{2}{s+1}$

i.e., $(s^2 + 4s + 13)\, L(y) = \dfrac{2}{s+1} - 1$

$$= \dfrac{1-s}{s+1}$$

$$L(y) = \dfrac{1-s}{(s+1)(s^2+4s+13)}$$

Now $\dfrac{1-s}{(s+1)(s2+4s+13)} = \dfrac{A}{s+1} + \dfrac{Bs+C}{s^2+4s+13}$

$$1 - s = A(s^2 + 4s + 13) + Bs + C(s + 1)$$

Put $s = -1$, therefore, $A = \dfrac{1}{5}$

Comparing the coefficient of s^2: $B = -\dfrac{1}{5}$

Comparing the coefficient of s: $C = -\dfrac{8}{5}$

Therefore, $\dfrac{1-s}{(s+1)(s^2+4s+13)} = \dfrac{1}{5}\left[\dfrac{1}{s+1}\right] + \dfrac{\left[\dfrac{-1}{5}\right]s + \left[\dfrac{-8}{5}\right]}{s^2+4s+13}$

Therefore, the solution is $y = \dfrac{1}{5}L^{-1}\left[\dfrac{1}{s+1}\right] - \dfrac{1}{5}L^{-1}\left[\dfrac{s+8}{s^2+4s+13}\right]$

$$= \dfrac{1}{5}e^{-x} - \dfrac{1}{5}L^{-1}\left[\dfrac{(s+2)+6}{(s+2)^2+3^2}\right]$$

$$= \dfrac{1}{5}e^{-x} - \dfrac{1}{5}e^{-2x}L^{-1}\left[\dfrac{s+6}{s^2+3^2}\right]$$

$$= \dfrac{1}{5}e^{-x} - \dfrac{1}{5}e^{-2x}(\cos 3x + 2\sin 3x).$$

2) Solve by Laplace Transform method $\dfrac{d^2y}{dx^2} + 3\dfrac{dy}{dx} + 2y = 2(t^2 + t + 1)$, given y(0) = 2, $y'(0) = 0$.

Sol: Given: $\dfrac{d^2y}{dx^2} + 3\dfrac{dy}{dx} + 2y = 2(t^2 + t + 1)$

Taking Laplace transforms on both sides,

L(y" + 3y' + 2y) = L[2(t² + t + 1)]

i.e., L(y") + 3L(y') + 2L(y) = 2L(t² + t + 1)

i.e., $[s^2 L(y) - sy(0) - y'(0)] + 3[s L(y) - y(0)] + 2 L(y) = 2\left[\dfrac{2}{s^3} + \dfrac{1}{s^2} + \dfrac{1}{s}\right]$

But, it is given that $y(0) = 2$, $y'(0) = 0$.

Therefore $[s^2 L(y) - 2s] + 3[s L(y) - 2] + 2 L(y) = =2\left[\dfrac{2}{s^3} + \dfrac{1}{s^2} + \dfrac{1}{s}\right]$

i.e., $(s^2 + 3s + 2) L(y) - (2s + 6) = =2\left[\dfrac{2}{s^3} + \dfrac{1}{s^2} + \dfrac{1}{s}\right]$

$$(s^2 + 3s + 2) L(y) = 2\left[\frac{2}{s^3} + \frac{1}{s^2} + \frac{1}{s}\right] + (2s + 6)$$

$$L(y) = \frac{2(s^2 + s + 2)}{s^3(s+1)(s+2)} + \frac{2s + 6}{(s+1)(s+2)}$$

$$(y) = L^{-1}\left[\frac{2(s^2 + s + 2)}{s^3(s+1)(s+2)}\right] + L^{-1}\left[\frac{2s + 6}{(s+1)(s+2)}\right]$$

Now, $\dfrac{2(s^2 + s + 2)}{s^3(s+1)(s+2)} = \dfrac{A}{(s+1)} + \dfrac{B}{(s+2)} + \dfrac{C}{s} + \dfrac{D}{s^2} + \dfrac{E}{s^3}$

$2(s^2 + s + 2) = A(s + 2)s^3 + B(s + 1)s^3 + C(s + 1)(s + 2)s^2 + D(s + 1)(s + 2)s + E(s + 1)(s + 2)$

Put $s = -1$, $A = -4$

Put $s = -2$, $B = 1$

Comparing the coefficents of s^2, $2 = 2C + 3D + E$ ➜ (1)

Comparing the coefficients of s, $2 = 2D + 3E$ ➜ (2)

Comparing the constant terms, $4 = 2E$ ➜ (3)

$\qquad$ (3) ➜ $E = 2$

Substituting $E = 2$ in (2), we get $D = -2$

Substituting $E = 2$ and $D = -2$ in (1), we get $C = 3$.

Therefore, $\dfrac{2\left(s^2 + s + 2\right)}{s^3(s+1)(s+2)} = -\dfrac{4}{(s+1)} + \dfrac{1}{(s+2)} + \dfrac{3}{s} - \dfrac{2}{s^2} + \dfrac{2}{s^3}$

Similarly, $\dfrac{2s+6}{(s+1)(s+2)} = \dfrac{A}{s+1} + \dfrac{B}{s+2}$

$2s + 6 = A(s + 2) + B(s + 1)$

Put s = – 1, A = 4

Put s = – 2, B = – 2

Therefore, $\dfrac{2s+6}{(s+1)(s+2)} = \dfrac{4}{s+1} - \dfrac{2}{s+2}$

Now, $y = L^{-1}\left[-\dfrac{4}{(s+1)} + \dfrac{1}{(s+2)} + \dfrac{3}{s} - \dfrac{2}{s^2} + \dfrac{2}{s^3}\right] + L^{-1}\left[\dfrac{4}{s+1} - \dfrac{2}{s+2}\right]$

$= -4e^t + e^{2t} + 3 - 4 + 6 + 4e^t - 2e^{2t}$

$= 5 - e^{-2t}).$

3)   Solve by Laplace Transform method $\dfrac{d^2y}{dt^2} + 6\dfrac{dy}{dx} + 5y = e^{-2t}$, given y(0) = 0, $\acute{y}$(0) = 1, when t = 0.

Sol: Given: $\dfrac{d^2y}{dt^2} + 6\dfrac{dy}{dx} + 5y = e^{-2t}$

Taking Laplace transforms on both sides,

L(y" + 6y' + 5y) = L(e^{-2t})

i.e., L(y") + 6L(y') + 5L(y) = L(e^{-2t})

i.e., [s^2 L(y) – sy(0) – y'(0)] + 6[s L(y) – y(0)] + 5 L(y) $= \dfrac{1}{s+2}$

But, it is given that y(0) = 0, y'(0) = 1.

Therefore [s^2 L(y) – 1] + 6s L(y) + 5 L(y) = $= \dfrac{1}{s+2}$

i.e., (s^2 + 6s + 5) L(y) $= \dfrac{1}{s+2} + 1$

$= \dfrac{s+3}{s+2}$

$$L(y) = \dfrac{s+3}{(s+2)(s^2+6s+5)}$$

$$L(y) = \frac{s+3}{(s+1)(s+2)(s+5)}$$

$$y = L^{-1}\left[\frac{s+3}{(s+1)(s+2)(s+5)}\right]$$

Now $\dfrac{s+3}{(s+1)(s+2)(s+5)} = \dfrac{A}{s+1} + \dfrac{B}{s+2} + \dfrac{C}{s+5}$

$S + 3 = A(s + 2)(s + 5) + B(s + 1)(s + 5) + C(s + 1)(s + 2)$

Put $s = -1$, therefore, $A = \dfrac{1}{2}$

Put $s = -2$, therefore, $B = -\dfrac{1}{3}$

Put $s = -5$, therefore, $C = -\dfrac{1}{6}$

Therefore, $\dfrac{s+3}{(s+1)(s+2)(s+5)} = \dfrac{\frac{1}{2}}{s+1} - \dfrac{\frac{1}{3}}{s+2} - \dfrac{\frac{1}{6}}{s+5}$

Therefore, the solution is $y = L^{-1}\left[\dfrac{\frac{1}{2}}{s+1} - \dfrac{\frac{1}{3}}{s+2} - \dfrac{\frac{1}{6}}{s+5}\right]$

$$= \frac{1}{2}L^{-1}\left[\frac{1}{s+1}\right] - \frac{1}{3}L^{-1}\left[\frac{1}{s+2}\right] - \frac{1}{6}L^{-1}\left[\frac{1}{s+5}\right]$$

$$y = \frac{1}{2}e^{-t} - \frac{1}{3}e^{-2t} - \frac{1}{6}e^{-5t}.$$

4) Solve by Laplace Transform method $\dfrac{d^2y}{dx^2} - 5\dfrac{dy}{dx} + 6y = e^{-x}$, given that $y = 0$, $x = 0$.

Sol: Given: $\dfrac{d^2y}{dx^2} - 5\dfrac{dy}{dx} + 6y = e^{-x}$

Taking Laplace transforms on both sides,

$$L(y'' - 5y' + 6y) = L(e^{-x})$$

i.e., $L(y'') - 5L(y') + 6L(y) = L(e^{-x})$

i.e., $[s^2 L(y) - sy(0) - y'(0)] - 5[s L(y) - y(0)] + 6 L(y) = \dfrac{1}{s+1}$

But, it is given that $y(0) = 0$, $y'(0) = 1$.

Therefore $[s^2 L(y) - 1] - 5s L(y) + 6 L(y) = \dfrac{1}{s+1}$

i.e., $(s^2 - 5s + 6) L(y) = \dfrac{1}{s+1} + 1$

$$= \dfrac{s+2}{s+1}$$

$$L(y) = \dfrac{s+2}{(s+2)(s^2 - 5s + 6)}$$

$$L(y) = \dfrac{s+2}{(s+1)(s-2)(s-3)}$$

$$y = L^{-1}\left[\dfrac{s+2}{(s+1)(s-2)(s-3)}\right]$$

Now $\dfrac{s+2}{(s+1)(s-2)(s-3)} = \dfrac{A}{s+1} + \dfrac{B}{s-2} + \dfrac{C}{s-3}$

$S + 2 = A(s - 2)(s - 3) + B(s + 1)(s - 3) + C(s + 1)(s - 2)$

Put $s = -1$, therefore, $A = \dfrac{1}{12}$

Put $s = 2$, therefore, $B = -\dfrac{4}{3}$

Put $s = 3$, therefore, $C = \dfrac{5}{4}$

Therefore, $\dfrac{s+2}{(s+1)(s-2)(s-3)} = \dfrac{\frac{1}{12}}{s+1} - \dfrac{\frac{4}{3}}{s-2} + \dfrac{\frac{5}{4}}{s-3}$

Therefore, the solution is $y = L^{-1}\left[\dfrac{\frac{1}{12}}{s+1} - \dfrac{\frac{4}{3}}{s-2} + \dfrac{\frac{5}{4}}{s-3}\right]$

$$= \dfrac{1}{12} L^{-1}\left[\dfrac{1}{s+1}\right] - \dfrac{4}{3} L^{-1}\left[\dfrac{1}{s-2}\right] + \dfrac{5}{4} L^{-1}\left[\dfrac{1}{s-3}\right]$$

$$y = \dfrac{1}{12} e^{-x} - \dfrac{4}{3} e^{2x} + \dfrac{5}{4} e^{3x}$$

VECTOR ANALYSIS

2 MARKS

1) Find $\nabla\varnothing$ if $\varnothing = x^2 + y^2 + z^2$.

   Sol: Given: $\varnothing = x^2 + y^2 + z^2$ ➜ (1)

   $$\nabla\varnothing = \frac{\partial\varnothing}{\partial x}\,\vec{i} + \frac{\partial\varnothing}{\partial y}\,\vec{J} + \frac{\partial\varnothing}{\partial z}\,\vec{k}$$

   From (1), $\frac{\partial\varnothing}{\partial x} = 2x, \frac{\partial\varnothing}{\partial y} = 2y, \frac{\partial\varnothing}{\partial z} = 2z$

   Therefore, $\nabla\varnothing = 2x\,\vec{i} + 2y\,\vec{j} + 2z\,\vec{k}$.

2) Show that the vector $= x^2z^2 + xyz^2 - xz^3$ is solenoidal.

   Sol: Given: $\vec{A} = x^2z^2\,\vec{i} + xyz^2\,\vec{J} - xz^3\,\vec{k}$

   $$\nabla\cdot\vec{A} = \frac{\partial}{\partial x}(x^2z^2) + \frac{\partial}{\partial y}(xyz^2) + \frac{\partial}{\partial z}(-xz^3)$$

   $$= 2xz^2 + xz^2 - 3xz^2 \qquad\qquad = 0$$

   Therefore, $\vec{A}$ is solenoidal.

3) State Green's Theorem.

   Sol: If U(x,y) and V(x,y) are two continuous functions of x and y having continuous partial derivatives $\frac{\partial U}{\partial y}$ and $\frac{\partial V}{\partial y}$ over aregion R bounded by a closed curve C in the xy – plane, then

$$\int U\, dx + V\, dy) = \iint \left[\frac{\partial V}{\partial x} - \frac{\partial U}{\partial y}\right] dx\, dy.$$

4) If $\varphi(x,y,z) = x^2y + y^2x + z^2$, find $\nabla\varphi$ at the point $(1,1,1)$.

Sol: Given: $\varphi(x,y,z) = x^2y + y^2x + z^2$ $\rightarrow$ (1)

$$\nabla\varphi = \frac{\partial\varphi}{\partial x}\vec{i} + \frac{\partial\varphi}{\partial y}\vec{J} + \frac{\partial\varphi}{\partial z}\vec{k}$$

From (1), $\dfrac{\partial\varphi}{\partial x} = 2xy, \dfrac{\partial\varphi}{\partial y} = 2yx, \dfrac{\partial\varphi}{\partial z} = 2z$

Therefore, $\nabla\varphi = 2xy\vec{i} + 2yx\vec{J} + 2z\vec{k}$

$$\nabla\varphi(1,1,1) = 2\vec{i} + 2\vec{J} + 2\vec{k} \qquad = 2\left(\vec{i} + \vec{J} + \vec{k}\right).$$

5) Find the divergence of $x^2\vec{i} + y^2\vec{J} + z^2\vec{k}$.

Sol: Let $A = x^2\vec{i} + y^2\vec{J} + z^2\vec{k}$

Divergence $\dfrac{\partial}{\partial x}(x^2) + \dfrac{\partial}{\partial y}(y^2) + \dfrac{\partial}{\partial z}(z^2) = 2(x + y + z).$

6) State Gauss Divergence Theorem.

Sol: If F is a continuously differentiable vector point function and V is the volume bounded by a closed surface S, then

$$\iint \vec{F}\,\hat{n}\, ds = \iiint div\left(\vec{F}\right) dV.$$

5 MARKS

1) If $\vec{F} = x^2 y\,\vec{i} + y^2 z\,\vec{J} + z^2 x\,\vec{k}$ then find curl(curl(F)).

Sol: Given: $\vec{F} = x^2 y\,\vec{i} + y^2 z\,\vec{J} + z^2 x\,\vec{k}$

$$\left(\text{Curl}\left(\vec{F}\right)\right) \nabla \times \left(\nabla \times \vec{F}\right)$$

Now, $(\nabla \times \vec{F}) = \begin{vmatrix} \vec{i} & \vec{J} & \vec{k} \\ \dfrac{\partial}{\partial x} & \dfrac{\partial}{\partial y} & \dfrac{\partial}{\partial z} \\ x^2 y & y^2 z & z^2 x \end{vmatrix}$

$$= \vec{i}\left[\frac{\partial}{\partial y}(z^2 x) - \frac{\partial}{\partial z}(y^2 z)\right] - \vec{J}\left[\frac{\partial}{\partial x}(z^2 x) - \frac{\partial}{\partial z}(x^2 y)\right] + \vec{k}\left[\frac{\partial}{\partial x}(y^2 z) - \frac{\partial}{\partial y}(x^2 y)\right]$$

$$= -y^2\,\vec{i} - z^2\,\vec{J} - x^2\,\vec{k}$$

Now, $\nabla \times (\nabla \times \vec{F}) = \begin{vmatrix} \vec{i} & \vec{J} & \vec{k} \\ \dfrac{\partial}{\partial x} & \dfrac{\partial}{\partial y} & \dfrac{\partial}{\partial z} \\ -y^2 & -z^2 & -x^2 \end{vmatrix}$

$$= \vec{i}\left[\frac{\partial}{\partial y}(-x^2) + \frac{\partial}{\partial z}(z^2)\right] - \vec{J}\left[\frac{\partial}{\partial x}(-x^2) + \frac{\partial}{\partial z}(y^2)\right] + \vec{k}\left[\frac{\partial}{\partial x}(-z^2) + \frac{\partial}{\partial y}(y^2)\right]$$

$$= 2z\,\vec{i} + 2x\,\vec{J} + 2y\,\vec{k}.$$

2) Using Green's Theorem, show that $\int (3x^2 - 8y^2)\,dx + (4y - 6xy)dy = 20$, where C is the boundary of the rectangle formed by the line x = 0, x = 1, y = 0, y = 2.

Sol: To show that $\int (3x^2 - 8y^2)\,dx + (4y - 6xy)dy = 20$

If is of the form $\int U dx + V dy$

$$U = 3x^2 - 8y^2, \quad V = 4y - 6xy$$

$$\frac{\partial U}{\partial y} = -16y, \quad \frac{\partial V}{\partial x} = -6y.$$

By Green's Theorem, $\displaystyle\int (U\,dx + V\,dy) = \iint \left[\frac{\partial V}{\partial x} - \frac{\partial U}{\partial y}\right] dx\,dy$

Therefore, $\displaystyle (3x^2 - 8y^2)dx + (4y - 6xy)dy = \int_0^2 \int_0^1 (-6y + 16y)\,dx\,dy$

$$= 10 \int_0^2 \int_0^1 y\,dx\,dy$$

$$= 10 \int_0^2 [xy]_{x=0}^{1}\,dy$$

$$= 10 \int_0^2 y\,dy$$

$$= 10 \left[\frac{y^2}{2}\right]_0^2 = 20.$$

3) If $r = x\,i + y\,j + z\,k$ and $|r| = r$, then show that $\nabla.(r^n\,r) = (n+3)r^n$.

Sol: $\operatorname{div}(r^n\,r) = . (r^n\,r)$

$$= r^n (. r) + r^n . r \;\longrightarrow \tag{1}$$

$$\left(\nabla.\varnothing A = \varnothing(\nabla.A) + \nabla\varnothing.A\right)$$

Now,

$$= \frac{\partial}{\partial x}(x) + \frac{\partial}{\partial y}(y) + \frac{\partial}{\partial z}(z) \qquad = 3$$

$$\nabla r^n = \sum \iota \frac{\partial}{\partial k}(r^n)$$

$$\sum inr^{n-1}\frac{\partial r}{\partial x}$$

$$\sum i\, nr^{n-1}\frac{x}{r}$$

$$= nr^{n-1}\sum xi$$

$$= nr^{n-1}\, r$$

Therefore, (1) → $div\ (r^n r) = 3r^n + nr^{n-2}\, r.r$

$$=3r^n+nr^{n-2}r^2$$

$$=3r^n+nr^n$$

$$=(3+n)r^n,$$

4) Find $\int F.dr$ where $\vec{F} = (x^2 - y^2)\hat{\imath} + 2xy\,\hat{\jmath}$ and C is the square bounded by the coordinate axes and the lines x = a, y = a.

Sol: Given: $\vec{F} = (x^2 - y^2)\hat{\imath} + 2xy\,\hat{\jmath}$

$$r = x\,\hat{i}+ y\,\hat{J}+ z\,\hat{k}$$

$$= x\,\hat{i}+ y\,\hat{J}\ \textit{(since z = 0 in xy – plane)}$$

$$\vec{dr} = dx\,\hat{i}+ dy\,\hat{J}$$

$$\vec{F}.\vec{dr} =[(x^2 - y^2)\hat{i}+2xy\,\hat{J}].(dx\,\hat{i}+ dy\,\hat{J})$$

$$= (x2 - y2)\ dx + 2\ xy\ dy\ → \qquad (1)$$

To find $\int \vec{F}.\vec{dr}$

C is the square OABD

(i)   Now consider

Along OA, y = 0. Therefore, dy = 0

Therefore, $\vec{F}.\vec{dr} = x^2 dx$

X varies from 0 to a

Therefore, along $OA, \int \vec{F}.\vec{dr} = \int_0^a x^2 dx = \left[\frac{x^3}{3}\right]_0^a = \frac{a^3}{3}$

(ii)   Along AB, x = a. Therefore, dx = 0

$F.dr = 2ay\,dy$

y varies from 0 to b.

Therefore, along

$OB, \int \vec{F}.\vec{dr} = \int_0^b 2ay\,dy = 2a\left[\frac{y^2}{2}\right]_0^b = 2a\left[\frac{b^2}{2}\right] = ab^2$

(iii) Along BD, y = b. Therefore, dy = 0

$\vec{F}.\vec{dr} = (x^2 - b^2)dx$

x varies from a to 0.

Therefore, along BD,

$BD, \int \vec{F}.\vec{dr} = \int_a^0 \left(x^2 - b^2\right)dx$

$= \left[\frac{x^3}{3} - b^2 x\right]_a^0$

$= ab^2 - \frac{a^3}{3}$

(iv)   Along DO, x = 0. Therefore, dx = 0

$$\vec{F}.\vec{dr} = 0$$

Therefore, along DO, $\int \vec{F}.\vec{dr} = 0$

Hence (1) ➜ $\int \vec{F}.\vec{dr} = 2ab^2$.

5)   Show that $\left[\dfrac{\vec{r}}{r^3}\right] = 0$, where $\vec{r} = x\,\hat{i} + y\,\hat{J} + z\,\hat{k}$

Sol: Given: $\vec{r} = x\,\hat{i} + y\,\hat{J} + z\,\hat{k}$

$$\text{div}\left[\frac{\vec{r}}{r^3}\right] = div\left(r^{-3}\,\vec{r}\right)$$

Since, $(\varnothing\,\vec{A}) = \varnothing(div\,\vec{A}) + \vec{A}.\nabla\varnothing,$

Therefore, $div\left[\dfrac{\vec{r}}{r^3}\right] = r^{-3}\left(div\,\vec{r}\right) + \vec{r}.\left(\nabla r^{-3}\right)$ ➜ (1)

Now, $div\,\vec{r} = 3$

$$\nabla r^{-3} = \hat{i}\,\frac{\partial}{\partial x}(r^{-3}) + \hat{J}\,\frac{\partial}{\partial y}(r^{-3}) + \hat{k}\,\frac{\partial}{\partial z}(r^{-3})$$

$$= \hat{i}\left[-3r^{-4}\,\frac{\partial r}{\partial x}\right] + \hat{J}\left[-3r^{-4}\,\frac{\partial r}{\partial y}\right] + \hat{k}\left[-3r^{-4}\,\frac{\partial r}{\partial z}\right]$$

$$= \hat{i}\left[-3r^{-4}\,\frac{x}{r}\right] + \left[-3r^{-4}\,\frac{y}{r}\right] + \left[-3r^{-4}\,\frac{z}{r}\right]$$

$$= -3r^{-5}(x\,\hat{i} + y\,\hat{J} + z\,\hat{k}) = -3r^{-5}\,\vec{r}$$

Therefore, (1) ➜ $div\left[\dfrac{\vec{r}}{r3}\right] = 3r^{-3} + \vec{r}.\left(-3r^{-5}\vec{r}\right)$

$$= 3r^{-3} - 3r^{-5}\,\vec{r}.\vec{r}$$

$$=3r^{-3} - 3r^{-5}\, r^2$$

$$=3r^{-3} - 3r^{-3} = 0.$$

6) Find the values of the constants a, b, c so that

$$\vec{F} = (x+2y+az)\hat{i} + (bx-3y-z)\hat{J} + (4x+cy+2z)\hat{k}$$

Sol: Given: $\vec{F} = (x+2y+az)\hat{i} + (bx-3y-z)\hat{J} + (4x+cy+2z)\hat{k}$

Therefore, $\nabla \times \vec{F} = 0$

$$\begin{vmatrix} \hat{i} & \hat{J} & \hat{k} \\ \dfrac{\partial}{\partial x} & \dfrac{\partial}{\partial y} & \dfrac{\partial}{\partial z} \\ x+2y+az & bx-3y-z & 4x+cy+2z \end{vmatrix} = 0$$

$$\hat{i}\left[\frac{\partial}{\partial y}(4x+cy+2z) - \frac{\partial}{\partial z}(bx-3y-az)\right]$$

$$-\hat{J}\left[\frac{\partial}{\partial x}(4x+cy+2z) - \frac{\partial}{\partial z}(x+2y+az)\right]$$

$$+\hat{k}\left[\frac{\partial}{\partial x}(bx-3y-z) - \frac{\partial}{\partial y}(x+2y+az)\right] = 0$$

Solving the above, we get a = 4, b = 2, c = – 1.

7) Evaluate by Green's Theorem $\int(xy + x^2)dx + (x^2+y^2)dy$, where C is the square

x = – 1, x = 1, y = – 1, y = 1.

Sol: By Green's Theorem,

Here U = xy + x^2 and V = x^2 + y^2

$$\frac{\partial U}{\partial y} = x; \frac{\partial V}{\partial x} = 2x$$

Given C is the square formed by the lines x = – 1, x = 1, y = – 1, y = 1.

To cover the whole square, x varies from to – 1 to + 1 and y varies from – 1 to + 1.

Therefore, $(xy + x^2)\,dx + (x^2 + y^2)\,dy = \int_{-1}^{1}\int_{-1}^{1}(2x - x)\,dx\,dy$

$$= \int_{-1}^{1}\int_{-1}^{1} x\,dx\,dy$$

$$= \int_{-1}^{1}\left[\frac{x^2}{2}\right]_{-1}^{1} dy$$

$$= \int_{-1}^{1}\left[\frac{1}{2} - \frac{1}{2}\right] dy$$

$$= \frac{1}{2}\int_{-1}^{1} 0\,dy \qquad = 0$$

10 MARKS

---

1) Verify Divergence Theorem for $\vec{F} = 4xz\,\hat{i} - y^2\,\hat{J} + yz\,\hat{k}$ takes over the cube bounded by x = 0, x = 1, y = 0, y = 1, z = 0, z = 1.

Sol: Given: $\vec{F} = 4xz\,\hat{i} - y^2\,\hat{J} + yz\,\hat{k}$

$$\nabla.\vec{F} = 4z - y$$

$$\iiint div\,\vec{F}\,dv = \int_{0}^{1}\int_{0}^{1}\int_{0}^{1}(4z - y)\,dx\,dy\,dz$$

$$= \int_0^1 \int_0^1 \left[2z^2 - yz \right]_0^1 dx\, dy$$

$$= \int_0^1 \int_0^1 (2 - y)\, dx\, dy$$

$$= \int_0^1 \left[2y - \frac{y^2}{2} \right]_0^1 dx$$

$$= \int_0^1 \left[2 - \frac{1}{2} \right] dx$$

$$= \int_0^1 \frac{3}{2} dx$$

$$= \left[\frac{3}{2}(x) \right]_0^1$$

$$= \frac{3}{2} \qquad \rightarrow \qquad (1)$$

To evaluate the surface integral, divide the closed surface S of the cube into faces.

S_1: The face AC'PB'

S_2: The face OBA'C

S_3: The face C'BA'P

S_4: The face OAB'C

S_5: The face B'PA'C

S_6: The face OAC'B

Therefore, $\iint \vec{F}.\hat{n}\,dS$ Double Integral Over $(S_1 + S_2 + S_3 + S_4 + S_5 + S_6)$

On S_1 we have x = 1. Therefore, $\vec{F} = 4z\,\vec{i} - y^2\,\vec{J} + yz\,\hat{k}$ and $\hat{n} = \vec{i}$

$$\vec{F}.\hat{n} = \left(4z\,\vec{i} - y^2\,\vec{J} + yz\,\vec{k} \right).\vec{i} = 4z$$

Elemental area on S_1 = dy dz

Therefore, Double Integral Over S_1 $\iint \vec{F}.\hat{n}\,dS = \int_0^1 \int_0^1 4z\,dy\,dz = 2$

On S_2 we have x = 0. Therefore, $\vec{F} = -y^2\,\vec{J} + yz\,\vec{k}$ and $\hat{n} = -\vec{i}$ $\vec{F}.\hat{n} = 0$

Therefore, Double Integral Over S_2 $\iint \vec{F}.\hat{n}\,dS = 0$

On S_3 we have y = 1. Therefore, $\vec{F} = 4xz\,\vec{i} - \vec{J} + z\,\hat{k}$ and $\hat{n} = \vec{J}$ $\vec{F}.\hat{n} = -1$

Elemental area on S_3 = dz dx

Therefore, Double Integral Over S_3 $\iint \vec{F}.\hat{n}\,dS = \iint -1\,dz\,dx = -1$

On S_4 we have y = 0. Therefore, and $\vec{F} = 4xz\,\vec{i}$ and $\hat{n} = \vec{J}$ $\vec{F}.\hat{n} = 0$

Therefore, Double Integral Over S_4 $\iint \vec{F}.\hat{n}\,dS = 0$

On S_5 we have z = 1. Therefore, $\vec{F} = 4x\,\vec{i} - y^2\,\vec{J} + y\,\hat{k}$ and $\hat{n} = \vec{k}$ $\vec{F}.\hat{n} = y$

Elemental area on S_5 = dx dy

Therefore, Double Integral Over S_5 $\iint \vec{F}.\hat{n}\,dS = \int_0^1 \int_0^1 y\,dx\,dy = \dfrac{1}{2}$

On S_6 we have z = 0. Therefore, $\vec{F} = -y^2\,\vec{J}$ and $\hat{n} = -\vec{k}$ $\vec{F}.\hat{n} = 0$

Therefore, Double Integral Over S_6 $\iint \vec{F}.\hat{n}\,dS = 0$

Hence, $\iint \vec{F}.\hat{n}\,dS = \dfrac{3}{2}$ $\Rightarrow$ (2)

From (1) and (2) we have $\iint \vec{F}.\hat{n}\,dS = \iiint div\,\vec{F}\,dV$

Hence the Divergence theorem is verified.

2) Verify Divergence Theorem for $\vec{F} = 4x\,\vec{i} - 2y2\,\vec{J} + z2\,\vec{k}$ over the region bounded by

$$x^2 + y^2 = 4,\ z = 0,\ z = 3.$$

Sol: Given: $\vec{F} = 4x\,\vec{I} - 2y2\,\vec{J} + z2\,\vec{K},\ z = 0, z = 3, x^2 + y^2 = 4.$

Let us have the following assumptions :

S_1: Bottom circular face in the plane $z = 0$.

S_2: Top circular face in the plane $z = 3$.

S_3: Curved surface.

On S_1: $z = 0$, $\hat{n} = -\vec{k}$

$$\vec{A}.\hat{n} = -\vec{A}.\vec{k} = -z^2 = 0$$

$$\iint \vec{A}.\hat{n}\,ds = \iint 0\,ds = 0 \quad\longrightarrow\quad (1)$$

On S_1: $z = 3$, $\hat{n} = \vec{k}$

$$\vec{A}.\vec{n} = \vec{A}.\vec{k} = z^2 = 9$$

$$\iint \vec{A}.\hat{n}\,ds = 9\iint ds = 36\pi \quad\longrightarrow\quad (2)$$

On S_3: $\hat{n} = \dfrac{\nabla x^2 + y^2 - 4)}{|\nabla x^2 + y^2 - 4)|} = \dfrac{2x\,\hat{i} + 2y\,\hat{J}}{2|x\,\hat{i} + y\,\hat{J}|}$

$$= \dfrac{x\,\hat{i} + y\,\hat{J}}{\sqrt{x^2 + y^2}} = \dfrac{x\,\hat{i} + y\,\hat{J}}{\sqrt{4}} = \dfrac{x\,\hat{i} + y\,\hat{J}}{2}$$

$$\vec{A}.\hat{n} = (4x\,\hat{\imath} - 2y^2\,\hat{J} + z2\,\hat{k}).\frac{x\,\hat{\imath} + y\,\hat{J}}{2} = 2x^2 - y^3$$

The cylindrical polar coordinates for S_3 are $x = r\cos\theta$, $y = r\sin\theta$, $z = z$, $ds = (r\,d\theta)\,dz$,

Where $r = 2$, $0 \le \theta \le 2\pi$, $0 \le z \le 3$ so that now

$$x = 2\cos\theta,\ y = 2\sin\theta,\ dS = 2\,d\theta\,dz.$$

$$\iint \vec{A}.\hat{n}\,ds = \int_{z=0}^{3} \int_{\theta=0}^{2\pi} \left[2(2\cos\theta)^2 - (2\sin\theta)^3 \right](2\,d\theta)\,dz$$

$$= 3\int_{\theta=0}^{2\pi} \left(16\cos^2\theta - 16\sin^3\theta \right) d\theta$$

$$= 48\int_{0}^{2\pi} (\cos^2\theta - \sin^3\theta)\,d\theta$$

$$= 48\left[\int_0^{2\pi} \frac{1+\cos 2\theta}{2}\,d\theta \right] - 48\left[\frac{\cos^3\theta}{3} - \cos\theta \right]$$

$$= 24\left[\theta + \frac{\sin 2\theta}{2} \right]_0^{2\pi} + 0$$

$$= 48\pi \quad\text{➜} \tag{3}$$

Therefore, $\iint \vec{A}.\hat{n}\,dS = 0 + 36\pi + 48\pi = 84\pi$ ➜ $\qquad(4)$

$$\iiint \nabla.\vec{A}\,dV = \iiint (4 - 4y + 2z)\,dV = 4\,\hat{x}V - 4\,\hat{y}V + 2\,\hat{z}V = 84\pi \text{ ➜ } (5)$$

Therefore, (4) = (5). Hence the Divergence theorem is verifed.

3)   Find the work done in moving a particle by a force $\vec{F} = 3xy\,\vec{i} - 5z\,\vec{J} + 10x\,\vec{k}$ along the curve $x = t^2 + 1$, $y = 2t^2$, $z = t^3$, from (2,2,1) to (5,8,8).

Sol: Given: $\vec{F} = 3xy\,\vec{i} - 5z\,\vec{J} + 10x\,\vec{k}$

The parametric equations of the curve are $x = t^2 + 1$, $y = 2t^2$, $z = t^3$

Evidently $t = 1$ gives (2,2,1) and $t = 2$ gives (5,8,8).

$$dx = 2t\,dt,\quad dy = 4t\,dt,\quad dz = 3t^2\,dt$$

work done $= \int \vec{F}.d\vec{r} \qquad = \int 3xy\,dx - 5z\,dy + 10x\,dz$

$$= \int_1^2 3\left(2t^4 + 2t^2\right)2t\,dt - \left(5t^3\right)4t\,dt + 10(t^2 + 1)3t^2\,dt$$

$$= \int_1^2 \left(12t^5 + 10t^4 + 12t^3 + 30t^2\right)dt$$

$$= \left(12\left[\frac{t^6}{6}\right] + 10\left[\frac{t^5}{5}\right] + 12\left[\frac{t^4}{4}\right] + 36\left[\frac{t^3}{3}\right]\right)_1^2$$

$$= ((128 + 64 + 48 + 80) - (2 + 2 + 3 + 10)) \qquad = 303.$$

4)   **Find the directional derivative of 2xy + 5yz + zx at the point (1,2,3) in the direction of the vector $3\,\vec{i} - 5\,\vec{J} + 4\,\vec{k}$.**

Sol: We know that the directional derivative in the direction of

$$\hat{e} = (\nabla\varnothing).\hat{e}$$

Here the given vector is $3\,\vec{i} - 5\,\vec{J} + 4\,\vec{k}$

The unit vector in the direction of the given vector is

$$\hat{e} = \frac{3\vec{\imath} - 5\vec{J} + 4\vec{k}}{\sqrt{3^2 + (-5)^2 + 4^2}} = \frac{3\vec{\imath} - 5\vec{J} + 4\vec{k}}{5\sqrt{2}}$$

Now,
$$\nabla\varnothing = \left[\vec{\imath}\,\frac{\partial}{\partial x} + \vec{J}\,\frac{\partial}{\partial y} + \vec{k}\,\frac{\partial}{\partial z}\right](2xy + 5yz + zx)$$

$$= \vec{\imath}\,(2y + z) + \vec{J}\,(2x + 5z) + \vec{k}\,(5y + x)$$

$$(\nabla\varnothing)_{(1,2,3)} = 7\,\vec{\imath} + 17\,\vec{J} + 11\,\vec{k}$$

So the directional derivative of $\varnothing$ at $(1,2,3)$ in the given direction is $\varnothing$ $(1,2,3)$

$$(\nabla\varnothing).\hat{e} = (7\,\vec{\imath} + 17\,\vec{J} + 11\,\vec{k}).\left[\frac{3\vec{\imath} - 5\vec{J} + 4\vec{k}}{5\sqrt{2}}\right]$$

$$= \frac{7(3) - 17(5) + 11(4)}{5\sqrt{2}} \qquad = \frac{21 - 85 + 44}{5\sqrt{2}}$$

$$= -\frac{20}{5\sqrt{2}} \qquad = -2\sqrt{2}.$$